Arthropods of the Lower Cambrian Chengjiang fauna, southwest China

HOU XIANGUANG AND JAN BERGSTRÖM

Hou X. & Bergström, J. 1997 12 22: Arthropods of the Lower Cambrian Chengjiang fauna, southwest China. *Fossils and Strata*, No. 45, pp. 1–116. Oslo. ISSN 0300-9491. ISBN 82-00-37693-1.

Arthropods from the Lower Cambrian Chengjiang fauna, Yunnan, southwest China, are described and compared with other arthropods, particularly with those of the Middle Cambrian Burgess Shale fauna. Where a direct comparison of taxa can be made, the new interpretations and reconstructions often differ from those based on the less well-preserved Burgess Shale fossils. Primitive arthropod features are represented in many species by, among other things, marked serial similarity in segments and podomeres, a large number of short podomeres in the endopod, a very small head, and stages of head accretion. The most primitive exopod appears to be a simple rounded flap. A slender exopod fringed with setae appears to be a key innovation shared by crustacean-like and trilobite-like arthropods. In the latter, the setae are large and flattened, and their setal character is clearly revealed by the basal articulations. The idea that the 'trilobite appendage' is restricted to trilobites is thus falsified. Within the dorsoventrally flattened trilobite-like arthropods, herein called lamellipedians, compound eyes tended to become dorsal and to become included in the head shield, and the appendages became laterally deflected. New taxa are as follows. Superclasses: Proschizoramia, Lamellipedia; classes and subclasses: Yunnanata, Paracrustacea, Megacheira, Artiopoda, Nectopleura, Conciliterga, and Petalopleura; orders and families: Fuxianhuiida and Chengjiangocarididae, Fortiforcipida with Fortiforcipidae, Acanthomeridiida with Acanthomeridiidae, Retifaciida with Retifaciidae, Skioldiidae and Saperiidae, Sinoburiida with Sinoburiidae, and Strabopiida; genera and species: *Fortiforceps foliosa* n. gen. et sp., *Squamacula clypeata* n. gen. et sp., *Kuamaia muricata* n. sp., *Skioldia aldna* n. gen. et sp., and *Almenia spinosa* n. gen. et sp. □*Arthropoda, Cambrian, China, Chengjiang fauna, taxonomy, evolution, Proschizoramia, Lamellipedia, Yunnanata, Paracrustacea, Megacheira, Artiopoda, Nectopleura, Conciliterga, Petalopleura, Fortiforcipida, Acanthomeridiida, Retifaciida, Sinoburiida, Strabopiida, Fuxianhuiidae, Chengjiangocarididae, Fortiforcipidae, Acanthomeridiidae, Retifaciidae, Skioldiidae, Saperiidae, Sinoburiidae.*

Hou Xianguang, Nanjing Institute of Geology and Palaeontology, Academia Sinica, Nanjing 210008, People's Republic of China; Jan Bergström [jan.bergstrom@nrm.se], Swedish Museum of Natural History, Box 50007, S-104 05 Stockholm, Sweden; 19th June, 1996; revised 22nd May, 1997.

Contents

Introduction

The Cambrian marks the apparent major radiation of Metazoa. For almost a century the Middle Cambrian Burgess Shale in British Columbia has been our main source of Cambrian animals with preservation of soft parts. These fossils have been significant for our understanding of the early evolution of the arthropods. Although we have been taught to marvel at the preservation of the Burgess Shale fossils, the fact is that strong compression and the toughness of the rock has presented difficulties in the study and interpretation of the fossils. Thus it is no wonder that specimens have already been reinterpreted a few times, and will no doubt continue to be so.

Later came new finds of well-preserved Cambrian fossils. By far the best quality is found in minute arthropods from bituminous limestone concretions (the *orsten*) in the Upper Cambrian Alum Shales of Sweden. These phosphatized remains, preserved without any compression, have been described by Klaus Müller and Dieter Walossek in a series of papers which revolutionize our understanding of arthropod evolution, particularly the evolution of early crustaceans and 'stem-group crustaceans'.

A second important fauna was found in the Lower Cambrian of the Kunming area in Yunnan Province, southwest China (Fig. 1) (e.g., Zhang & Hou 1985; Hou & Sun 1988; Hou *et al.* 1991), and is partly described here. Despite the difference in age, it is taxonomically very similar to the Burgess Shale fauna. The preservation is much different, and it is easier to prepare and study the specimens. This material therefore promises to solve quite a number of the problems caused by insufficient preservation of Burgess Shale material and to pose a number of new questions. One important aspect is the age of the Chengjiang fauna; it is middle Lower Cambrian (lower part of the *Eoredlichia* Zone) and thus quite close to the base of the Phanerozoic, where animals first appear as fossils (Figs. 3, 4). The advanced state of the life forms so low down in the sequence of sedimentary rocks is startling, but there are also individual forms that appear primeval and may guide us in our search for an origin of animal groups, particularly arthropods.

In order to evaluate the systematic and evolutionary position of the Chengjiang arthropods, we have had consider a number of other Cambrian and some Devonian arthropods. This could not be done simply by referring to the literature, since important structures have often not been described, or have been misinterpreted.

The authors share responsibility for the entire text even though HXg had more influence on the descriptive part, JB more on the discussion.

Previous geological research in the Kunming area

The study of the Lower Cambrian in the Kunming area, eastern Yunnan (Fig. 1A), has a long history. As early as in 1909 and 1910 the French scientists J. Deprat and H. Mansuy studied the geology and palaeontology of this area (e.g., Mansuy 1912). In the 1930s and 1940s, the Lower Cambrian stratigraphy and phosphorites of the Kunming area were extensively studied (e.g., Wang H.-z. 1941; Wang Y.-l. 1941; Ho 1942). During his investigation of phosphorite reserves, Ho (1942) measured the section at Maotianshan in Chengjiang County (Fig. 1B) and introduced the term 'Maotianshan shale system' for the Lower Cambrian mudstone we now know to contain the Chengjiang fauna. Lu (1941) systematically studied the Lower Cambrian stratigraphy and trilobites of this district and named the Lower Cambrian Chiungchussu (Qiongzhusi), Tsanglangpu (Canglangpu) and Lungwangmiao (Longwangmiao) Formations. The Lower Cambrian of the Kunming area in eastern Yunnan has long been taken as a standard for stratigraphical subdivision and correlation within the southwest China Platform and also for all China and the Redlichiid realm as a whole.

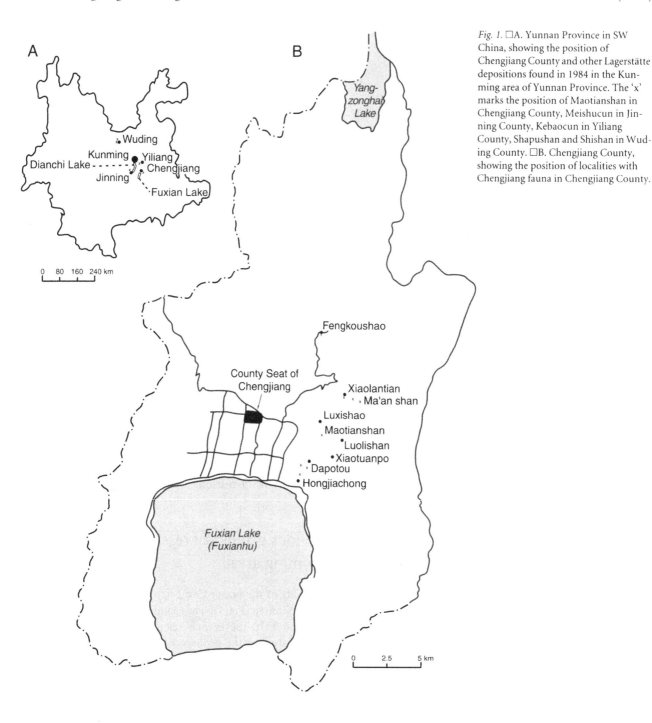

Fig. 1. □A. Yunnan Province in SW China, showing the position of Chengjiang County and other Lagerstätte depositions found in 1984 in the Kunming area of Yunnan Province. The 'x' marks the position of Maotianshan in Chengjiang County, Meishucun in Jinning County, Kebaocun in Yiliang County, Shapushan and Shishan in Wuding County. □B. Chengjiang County, showing the position of localities with Chengjiang fauna in Chengjiang County.

Discovery and study of the Chengjiang fauna

The discovery of the soft-bodied Chengjiang fauna was the result of extensive and intense field work in the area by one of us (HXg). The initial reason for the field work was HXg's research program involving the study of bradoriid arthropods, which were known to occur abundantly in the Lower Cambrian Qiongzhusi (Chiungchussu) Formation on the southwest China (or Yangtze) Platform (e.g., Huo 1956; Lu, Yu & Chen 1981) and

which required fundamental taxonomic revision (Hou *et al.* 1996; Hou 1997).

During the 1930s, Zhongshan University had its Department of Geology moved to Donglongtan, Chengjiang, about 1 km west of Maotianshan (Maotian Hill), the hill where HXg first discovered (in 1984) soft-bodied fossils of the Chengjiang fauna. The Lower Cambrian shale near the Zhongshan University is rich in fossils identified as *Bradoria* sp. by Yang Zui-yi, the then Professor of Geology of Zhongshan University (Ho 1942; Yang Z.-y. 1991, personal communication).

Fig. 2. Maotianshan (Maotian Hill) towards the east in July 1984.

HXg began the journey from Nanjing to Yunnan Province on 5th June, 1984, and Kunming was reached on the 9th. After a preliminary investigation in Jinning County, HXg arrived at the nominal town of Chengjiang County, 56 km southeast of Kunming, on 19th June. A geological team, the Seventh Division of the First Geological Group, Geological Bureau of Yunnan Province, was prospecting for Lower Cambrian phosphorite. The following day HXg travelled to their camp in Dapotou, a small village situated in the upland (Fig. 1B), and was welcomed by the leaders of the team.

After four days' investigation of the Qiongzhusi Formation at Dapotou, Hongjiachong, Xiaotuanpo, Maotianshan, Luolishan, and nearby areas (Fig. 1B), HXg selected a section near Hongjiachong village for the systematic collection of bradoriids because of its well-exposed and seemingly undisturbed sequence. A local peasant was hired to assist in digging the mudstone. Work proceeded for five days, during which time water and food had to be brought to the hill. The mudstone in the section was only slightly weathered, which made work difficult. Closer inspection of the locality indicated the presence of a gap in the sequence, probably caused by a fault; this caused search for an alternative section.

The following day, Sunday July 1st, HXg and the peasant walked to Maotianshan (Fig. 2), a place already vis-ited a few days earlier by a different route. A section on the west slope of Maotianshan was finally selected for study, and the peasant dug out mudstones, which were split for bradoriids. Work was notably easier than at Dapotou and Hongjiachong, because the rock was strongly weathered. By about three o'clock in the afternoon, a semi-circular white film on a slab turned up; at the time it was mistaken for an unknown crustacean valve, with one straight and one curved side. The second find was of an exoskeleton with subelliptical shape. The third find was most revealing. On splitting a slab, a complete animal, some 4–5 cm long, turned up. The two immediately previous finds proved to be merely two dorsal tergites of the entire animal. Partial stripping of the tergites revealed segmentally arranged imbricating limbs. The midline had some similarity with the backbone of a vertebrate. The animal almost appeared alive on the wet surface. This specimen was ultimately selected as the holotype of *Naraoia longicaudata* (Zhang & Hou 1985). That day, work in the section did not end until almost dark, although the walk back to Dapotou would take an additional hour. The find reminded HXg of the Burgess Shale fauna; his field diary notes the important find: 'The discovery of fossils in the Phyllopod Bed', and it was difficult for him to sleep that night.

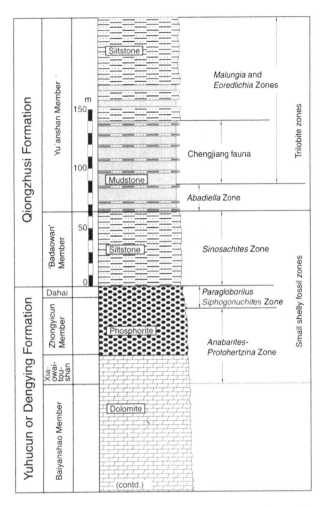

Fig. 3. The sequence near the Precambrian–Cambrian boundary in the Kunming area. It also is taken as a standard for stratigraphical subdivision and correlation within the Southwest China Platform, and for all China and the Redlichiid realm as a whole.

The Geological Team assisted by blasting the west slope of Maotianshan. Soft-bodied and other fossils were then collected from three more or less comprehensive levels in the west slope, designated M2 (oldest), M3 and M4 (youngest), respectively. These designations were subsequently used for three fossil quarries opened in the fossiliferous levels. The three levels do not correspond to three but to over ten beds with soft-body preservation (Hou 1987a). It is in fact almost impossible to determine exactly how many beds, from bottom to top, in the blocky mudstone that bear soft-bodied fossils. The mudstone of level M2 is 5 m thick and yields almost all elements of the Chengjiang fauna, while upwards through levels M3 and M4 both the number of taxa and specimens successively decrease and may reflect successive deterioration of conditions of preservation.

Systematic search for bradoriids was also performed at several localities in eastern Yunnan (Fig. 1A): at Meishu-

cun in Jinning; at Sapushan (Sapu Hill) and Shishan (Shi Hill) in Wuding; at Kebaocun in Yiliang as well as at Hongjiachong, Dapotou and Maotianshan in Chengjiang. In addition to bradoriids, a few specimens of *Isoxys* and worms were also collected from the sections at Meishucun (Chinese candidate for a global stratotype for the Precambrian–Cambrian boundary), Sapushan and Kebaocun. Especially significant were discoveries of two specimens of *Heliomedusa* and one specimen of *Cricocosmia* in the Meishucun section and one isolated sclerite of *Microdictyon* in the Shishan section. Additional well-preserved specimens were collected in the Meishucun section on 31 August, 1986 (see Hou & Sun 1988). As a result of extensive field work, it appears that fossils with soft-part preservation are widely distributed in eastern Yunnan, but they are rare. Thus, splitting of a large amount of rock is necessary for the collection of a reasonable number of specimens.

HXg's field work ended after ten weeks, on 17th August, 1984. Letters sent by HXg from the field in 1984 reported to the directors and others at the Institute of Palaeontology in Nanjing three main results of his work: (1) The discovery and collection of many fossils with preserved soft parts; (2) the collection of a large number of well-preserved bradoriid specimens (some of which were treated by Hou 1987d); and (3) the discovery and collection of the oldest trilobites at Chengjiang, Wuding and Jinning. Some of the latter were reported by Zhang (1987a). Other trilobites collected at Maotianshan by HXg were reported separately by Zhang (1987b).

The goal of a second field trip, from 12th April to 10th June, 1985, was to collect fossils with soft-part preservation. Logistics had changed: the Geological Team in Dapotou had new leaders; the ground had been cleared on both sides of the cart road leading to Maotianshan (for two new phosphorite factories); lastly, a drilling team (Group 809, Geological Bureau of the Yunnan Province) was living at the foot of Maotianshan (Fig. 2) and was very helpful, which included their providing us with lunch and water.

	China	Eustatic events	Biological events
Lower Cambrian	Qiongzhusi	Transgression events	Radiation of Arthropoda
		Regression events	
	Meishucun	Phosphogenesis and transgression event	Radiation of 'small shelly' fauna
UPZ	Dengying	Regression events	Vendobionts

Fig. 4. Interpretation of eustatic and biological events in the latest Proterozoic and Early Cambrian.

With the support of the directors of the Nanjing Institute of Geology and Palaeontology, Academia Sinica, fieldwork continued for 82 days in October to December of 1985, during which period the first phosphate plant at Dapotou started production and Chen Luansheng joined Hou Xianguang for some time. Large-scale collecting took place from April to September, 1987, mainly at Maotianshan and Jianbaobaoshan near Dapotou, again with the support of Academia Sinica. Chen Jun-yuan, Zhou Gun-qin, and Zhang Jun-ming were added to the field group but left the field early in May or early in June for other duties. Additional collections were made by Hou Xianguang in November, 1989, and in April and May, 1990; this work included the localities Ma'anshan, Fengkoushao and Xiaolantian (Fig. 1B).

Initially a number of short descriptions of fossils from the Chengjiang fauna were published, the first one by Zhang & Hou (1985). In 1989, cooperation between HXg and Swedish scientists was arranged by Academia Sinica and the Institute in Nanjing. Subsequently, several individuals and groups have appropriated parts of the fauna.

Stratigraphy and localities

The Lower Cambrian of eastern Yunnan is divided into 14 biozones (Fig. 5). The first three are based on small shelly fossils and the following eleven on trilobites (e.g., Zhou & Yuan 1982). The oldest trilobite genus, *Abadiella*, characterizes the fourth biozone. It occurs in eastern Yunnan in a 3 m interval of black siltstone and mudstone with dolomitic and pyritic concretions. The base of this level is generally taken as the base of the Yu'anshan Member of the Lower Cambrian Qiongzhusi (Chiungchussu) Formation and the beginning of the Qiongzhusian Stage. The succeeding biozone is typified by the trilobite genera *Eoredlichia* and *Wutingaspis*. In eastern Yunnan the fourth and fifth zones are separated by a 10–30 m thick sequence without the zonal fossils. This is the case at Maotianshan, where the interval is 30 m thick. Here the Chengjiang lagerstätte has its base less than 20 m above the *Abadiella* bed, which is about 1 m thick (Hou 1987a).

With the exception of those specimens from the Meishucun section in Jinning County, the soft-bodied

	Lithostratigraphy		Age(Ma)	Biozone	
Lower Cambrian	Lungwangmiao Formation			14. *Redlichia guizhouensis*	Trilobite zones
				13. *Hoffetella* Zone	
	Tsanglangpu Formation			12. *Megapalaeolenus* Zone	
				7. *Yunnanaspis – Yiliangella* Zone	
	Chingchussu (Qiongzhusi) Formation	Yu´anshan Member		6. *Malungia* Zone	
				5. *Eoredlichia* Zone	
				Chengjiang fauna	
				4. *Abadiella* Zone	
		'Badaowan' Member	Rb 579 Rb 587	3. *Sinosachites – Eonovitatus* Zone	Small shelly fossil zones
	Yuhucun (Dengying) Fm.	Dahai Mbr.		2. *Paragloborilus – Siphogonuchites* Zone	
		Zhongyicun Member			
		Xiaowaitoushan Mbr.		1. *Anabarites – Circotheca* Zone	
?					
Proterozoic		Baiyanshao Member			

Fig. 5. Lithostratigraphic units and biozonation of the Lower Cambrian in Southwest China. The Precambrian–Cambrian boundary is supposed to be below the *Anabarites* and *Circotheca* Zone, whereas Zones 1–14 belong to the Lower Cambrian. The Meishucun Stage corresponds to the small shelly fossil zones 1–3, the Chiungchussu (Qiongzhusi) Stage to Zones 4–6 (note the different scope of the Chiungchussu Formation!), the Tsanglangpu (Canglangpu) and Lungwangmiao (Longwangmiao) Stages to the formations with the same name.

Yunnan	Acritarch zones	Balto- scandia	East Baltic	Siberia	age Ma
Longwangmiaoan Canglangpuan Qiongzhusian*	*Volkovia dentifera*	'Protolenus'	Rausve	Toyonian Botomian Atdabanian	525
	Micrhystridium dissimilare	*Holmia kjerulfi*	Vergale	U. Tommotian M. Tommotian	
Meishucunian	*Skiagia ornata*	*Schmidtiellus mickwitzi*	Talsy	L. Tommotian	530
	Comasphaerid. velvetum & *A. tornatum*	*Platysolenites antiquissimus*	Lontova	Yudomian	540

Fig. 6. Correlation of the Lower Cambrian. An asterisk denotes the position of the Chengjiang fauna. Siberian stage correlation according to Moczydłowska & Vidal (1988) and Palacios & Vidal (1992).

fossils collected in Yunnan province are from localities in Chengjiang County (Fig. 1B):

1 The west slope of Maotianshan, levels M2, M3 and M4 (in ascending order).

2 The northwestern slope of Maotianshan, levels Cf1 to Cf8 (in ascending order). Cf1 through Cf4 are along a road. Cf1 and Cf2 are below the level of M2; no trilobites have been found. Cf3 through Cf6 correspond to levels M2 and M3, Cf7 and Cf8 to M4.

3 The eastern side of Jianbaobaoshan, where the mudstone is strongly weathered. The locality is about 300 m west of Dapotou village. Levels Dj1 and Dj2, which correspond to level M2 of Maotianshan.

4 The Xiaolantian section, about 500 m southeast of Xiaolantian village. Level XL1, which corresponds to level M2.

5 The Ma'anshan section, about 1 km southeast of Xiaolantian village and 3 km northeast of Maotianshan. Level Ma1, which corresponds to level M2.

6 The Fengkoushao section, in the Fengkoushao village. Level FK1, which roughly corresponds to level M2.

7 The Meishucun section (Jinning County); this was a candidate stratotype section for the Precambrian–Cambrian boundary. Level K10, located 55 m above the *Abadiella* beds, which corresponds to level M2 of Maotianshan.

Correlation

Correlation in the Lower Cambrian has been unusually difficult, both because of the relative rarity of fossils and because of the provinciality of the faunas. The custom of correlating the boundary between 'pre-trilobite' and 'trilobite' strata, or correlating with 'small shelly fossils', is now slowly giving way to the use of less habitat-dependent evidence, including the vertical distribution of acritarchs.

Studies on acritarchs in the southeast Yunnan sequence (Zang 1992) allows a tentative correlation of the boundary between the *Skiagia ornata* and the *Micrhystridium dissimilare* acritarch Zones with the Meishucunian–Qiongzhusian boundary (Fig. 6). This level is probably close to the top of the Tommotian (e.g., Moczydłowska & Vidal 1988; Palacios & Vidal 1992), so far as this 'horizon' is defined in Siberia. This corresponds roughly to the boundary between the Talsy and Vergale Formations in the East Baltic. The way that correlation was made in the former Soviet Union misled us to believe that this boundary was in the middle Atdabanian rather than within or close to the top of the Tommotian, which was one reason for misunderstanding the 'Siberian' use of terms. In terms of Baltoscandian stratigraphy, this level corresponds to a level in the middle of the sequence with *Holmia*, between the zones with *Schmidtiellus mickwitzi* and *Holmia kjerulfi*. The level of the Chengjiang fauna may thus correspond roughly to the *Holmia kjerulfi* Zone. This appears to be Atdabanian in Siberian terms.

Material and methods

The Chengjiang fossils are preserved in a soft mudstone. This and the fact that the fossils are not completely flat make it possible to prepare the specimens carefully with a needle.

For the discussion of aspects of evolution, phylogeny and systematics, it was necessary to include information from arthropods other than those from Yunnan. This includes trilobites as well as arthropods from other lagerstätten, such as the *orsten* concretions, the Burgess Shale and the Hunsrück Slate. Some text is therefore included on such arthropods, giving reinterpretations where necessary. The appendages of bradoriids are dealt with in a separate paper (Hou *et al.* 1996).

The catalogue number in the collections of Academia Sinica, Nanjing, is indicated by the letter combination

CN (illustrated specimens only). The specimens are housed in the Museum of the Nanjing Institute of Geology and Palaeontology, Academia Sinica, Nanjing. In our text, page references for publications in Chinese refer to the English summary, but in the reference list they refer to the entire paper.

Mode of preservation

The Chengjiang fossils include individually preserved carapaces and whole individuals with variously preserved soft parts. Particularly in the case of anomalocaridids, there are remains that appear to have been in a stage of decay before being embedded in sediment. On the whole, however, there are a surprising number of complete animals, such as arthropods, diverse types of worms, and also brachiopods with setae and pedicles preserved. It appears that organic material has been comparatively well preserved. Since the colour darkens with the thickness of the skeleton, we believe that it is in part the organic material that gives the skeletal and soft-part remains a different colour from that of the host mudstone. Occasionally there is also an indication of mineral colouring. There is no evidence of preservation of carbonate in our material; however, this is possibly because most of the material collected is fairly weathered.

Appendages are commonly visible through the body. Each appendage is then often expressed not as a ridge, but as a shallow furrow. The reason for this apparent paradox is that the appendage has collapsed, leaving an empty space into which the dorsal exoskeleton has been pressed (cf. Chen et al. 1995b, p. 276).

The remains are flattened, but there is a distinctive difference in the degree of flattening between the Burgess Shale and the Chengjiang fossils. While the Burgess Shale specimens are paper-thin, a certain relief is seen among the Chengjiang fossils. Thus, for instance, the exopod setae often show no evidence of flattening (Figs. 22, 25, 39, 41). Instead, they present a thin edge towards the viewer, the flat sides sloping downward at an angle with the horizontal (e.g., about 60° in *Naraoia*). Often it is also possible to see in low-angle light a difference in slope between left and right sides. This degree of relief facilitates preparation, particularly in combination with a degree of weathering.

Terminology

For the lamellipedians, the set of terms used for trilobites (Moore 1959) is generally applicable. However, the more general terms *head* and *tail* are used rather than *cephalon* and *pygidium*. For other arthropods the terminology used

for crustaceans (Moore & McCormick 1969) and to some degree for chelicerates (Størmer 1955) has been used.

Walossek & Müller (1990) showed that the coxa is a secondarily formed proximal segment of the biramous leg in crustaceans. It does not exist in 'stem-lineage crustaceans' and apparently not in trilobite-like arthropods, the proximal leg segment of which obviously corresponds to the crustacean basis. This reinterpretation means that the two branches of the trilobite limb correspond to those in the crustacean limb and can be called *endopod* and *exopod*. The proximal articulation and the stiffness of the exopod lamellae ('gill filaments') indicate that they are setae, closely comparable with correspondingly positioned setae in crustaceans. This interpretation (e.g., Bergström 1992, p. 290) is accepted by Ramsköld & Edgecombe (1996). An anterior plate in the head of helmetiids, separated from the rest of the head by a suture, is referred to as a *rostral plate*. This sclerite compares with the 'head' shield in *Sidneyia*, which appears to cover a minihead consisting of only acron and antennal segment, and at the same time has a large doublure, which is similar in position to the hypostome of trilobites. The hypostome is defined as a medial ventral exoskeletal cover extending to the region of the mouth. It can be an expansion of the cephalic doublure or an isolated sclerite.

Acron: Pre-segmental part of arthropod body, in front of the antennal segment.

Annulus: Ring of an antenna.

Antenna: The anteriormost appendage, usually uniramous. In crustaceans and megacheirans, and probably also in the trilobite *Kuanyangia*, the 2nd appendage is also not developed as a leg, but it is comparable with the 2nd antenna of crustaceans. In this case, it is convenient to differentiate between 1st and 2nd antenna. In crustacean literature, these are commonly called antennula and antenna, respectively (the antennula thus corresponds to the antenna of other arthropods).

Basis: The proximal podomere in primitive schizoramian arthropods.

Carapace: The definition of this term has been very vague and troublesome when applied to crustaceans. As used herein, it simply means a free tergal prolongation backwards from the head of an arthropod, whichever segment(s) it may arise from and without reference to homology.

Counterpart: This term is used in descriptions of the Burgess Shale fauna. We use the more descriptive expression 'upper part'.

Coxa: Secondarily formed podomere of appendage, proximal to the basis; apparently absent from schizoramian arthropods other than crustaceans.

Downward view: View of the lower part of a horizontally cleft specimen.

Endopod: Inner branch of biramous leg, arising from the basis.

Exopod: Outer branch of biramous leg, arising from the basis.

Facial suture: Typically, line along which head tergite separates into pieces during moulting (as in most trilobites); may be non-functional.

Filament: In the terminology used for Burgess Shale arthropods, a filament is a long seta or spine (cf. 'gill').

Gill: In the terminology used for Burgess Shale arthropods, a gill is any structure carrying notably long setae or spines, whether associated with legs or with dorsal tergites. It is also used for the exopod of the limb even when it is devoid of setae. The term is not used herein.

Hypostome: Ventral sclerite situated in front of the mouth and separate from the rostral plate or doublure.

Lower part: This corresponds to the term 'part' that is used in the descriptions of the Burgess Shale fossils. For a dorsoventrally flattened fossil, it means the lower part of the two that come out of a horizontal split. This part is seen in dorsal view, but as the split occurs through the fossil itself, it often does not show the most external dorsal parts. The complementary term is 'upper part' (='counterpart' in descriptions of the Burgess Shale fauna). For the direction of view, the expressions 'downward view' and 'upward view' are used.

Pararostral plate: Paired plate at the anterior margin of the head, lateral to the rostral plate.

Part: Term used in the descriptions of Burgess Shale fossils. We use the more descriptive expression 'lower part'. See 'downward view' and 'lower part'.

Podomere: Segment of leg.

Pseudosegmentation: False segmentation; state in which there is an uncoordinated repetition of individual organs but no repetition of serially similar body compartments.

Rostral plate: Anterior portion of head shield on ventral and dorsal sides. It is demarcated posteriorly by a transverse facial suture and laterally by sutures against the pararostral plates. It may, in cases, be prolonged ventrally into a hypostome-like extension.

Segmentation: Repetition of body divisions, each typically or at least primitively having a separate sclerite ring and one set each of limbs and different internal organs. Also repetition of limb divisions.

Semitergite: Tergite that is more or less separate axially but fused to adjacent tergites laterally.

Seta: Hair-like or spine-like process of cuticle with which it is articulated.

Sternite: Skeletal plate on the ventral side of the animal.

Telosoma: Posterior body portion lacking legs.

Telson: Postsegmental segment of arthropod body; gives rise to new segments during growth.

Tergite: Skeletal plate on the dorsal side of the animal.

Upper part: The upper part of a horizontally cleft specimen.

Upward view: View of the upper part of a horizontally cleft specimen.

Some arthropod features

Key characters of arthropods are an exoskeleton and a segmented body, segments of which carry segmented appendages.

The arthropod skeleton is an exoskeleton and consists of the cuticle (Neville 1975). The cuticle is layered, the two basic parts being the epicuticle and procuticle. In many arthropods with a hard exoskeleton, the procuticle is subdivided into exocuticle, mesocuticle and endocuticle. Three to four types of chemical components are involved in building up the cuticle. Of these, chitin is a cellulose-like polysaccharid, which is water-soluble and absent from the exposed epicuticle. A variety of lipids, such as hydrocarbons, wax esters, cuticulin, alcohols, fatty acids and sterols, stabilise the cuticle and make the epicuticle impenetrable to water. Proteins are the main structural component of cuticles. Hardening of the cuticle is performed through tanning of the proteins in the exocuticle (sclerotization), or through impregnation with minerals (commonly calcium carbonate but also, for instance, hydroxyapatite). Hard parts are called sclerites and are separated by soft and flexible arthrodial membranes. Sclerotized mandibles of insects may achieve a hardness of 3 on Moh's scale for minerals. Calanoid copepod crustaceans, feeding on hard diatoms, have their epicuticular mandibular teeth impregnated with opal, which gives them a hardness of 5.5 to 6.5 (i.e. intermediate between apatite and quartz). At the other end of the scale is the soft and penetrable cuticle of gills.

The cuticle cannot expand. When the animal grows, it therefore has to undergo moulting. Thus, the cuticle is shed (in a process called ecdysis), and a new cuticle is produced by a sheet of epidermal cells underlying the cuticle.

The animal is usually divided into tagmata with typical specializations. There is typically a head with modifications for orientation and treatment of food. The body behind the head may be uniform or divided into tagmata that can be specialized for, e.g., walking and swimming. The appendages are often modified for a variety of functions.

The appendages described herein lack the proximal podomere (coxa), which at one time was considered primitive for all arthropods. The most proximal podomere is the basis. This statement is a logical consequence of the important revelation of how the coxa first developed in the phylogenetic lineage leading to the crustaceans (Walossek & Müller 1990). The absence of a coxa, not to say of a precoxa, in trilobites and other early arthropods makes all previous comparisons with the crustacean appendage incorrect. Two branches extend from the basis. These are the endopod and, outside or above it, the exopod. Because of the erroneous older identification of the basis as a coxa or precoxa, the exopod was supposed to be an exite and to have had the respiratory function of crustacean exites. In fact, its function varies from one group to the next. In the body region, the endopod is most commonly used for walking. When this is the case in crustaceans and merostomes, the exopod is often lost. In other cases both branches are used for swimming or for filter-feeding.

In crustaceans, much of the respiration is through the integument of the carapace or pleural underside; for example, through the inner lamella in most ostracodes. This may also have been the case in *Agnostus* (Müller & Walossek 1987, p. 38). Similarly, respiration in trilobites was probably through the ventral integument of the pleurae. In crustaceans, additional particular gills may be formed by particular soft coxal and precoxal outgrowths called epipodites. No similar gills have been identified in any Palaeozoic arthropods, except perhaps for club-like outgrowths on the endopodal podomeres in *Agnostus* (Müller & Walossek 1987, pp. 37–38, Figs. 4, 6). In many branchiopod crustaceans, the entire limb is flat and its thin cuticle allows gas exchange. It functions as a gill as well as serving in feeding and locomotion. Similarly, the swimming appendages or pleopods of isopod crustaceans function as gills. Feeding is often facilitated by enditic outgrowths, which may have spines on the inside or underside of the endopods.

Setae are a group of outgrowths on the limbs that may be superficially similar to hairs or spines. A seta differs from a spine in being attached by a joint. Setae are used for many functions by crustaceans. In many groups, such as the Remipedia, Anaspidacea, Tanaidacea, Cumacea, Euphausiacea, and Reptantia, setae are used in swimming by expanding the surface of the swimming appendages, which are often concentrated on the abdomen. In cephalocarids, leptostracans, most branchiopods, and nauplius larvae, setae are involved in 'filter' feeding. In stomatopods, the setose 1st maxilliped is used for grooming and for cleaning the antennae, eyes and mouthparts. In many cases the exact function of the setae is not understood but, on the whole, in Recent arthropods they have a mechanical function, either in interacting with the surrounding water, in dealing with the food, or in cleansing the body. The function of setae in extinct arthropods is obviously not directly observable, but trace fossils produced by trilobite-type appendages demonstrate that in some species the exopod setae were used to stir up the sediment of the sea-bottom (Bergström 1976b).

Taxonomy and systematics

The taxonomic rank of groups is a difficult problem to resolve. Manton (1978) held that the Crustacea were a discrete phylum. There has been a general elevation of the rank of many arthropod groups so that, for instance, the Class Crustacea has given way to some 12 classes of extant crustaceans in most modern classifications (e.g., Storch & Welsch 1991). If we follow this philosophy, most crustaceans and crustacean-like arthropods in the Upper Cambrian *orsten* will represent new monotypic classes, and the 'stem-group crustaceans' of Walossek & Müller (1990) will belong to a different phylum, or to several different phyla if cladistic principles are adhered to strictly. Still more phyla would be needed for more distantly related arthropods.

The elevation of ranks is thus, in part, a result of ignorance of extinct groups. In order to resolve this dilemma, we have adopted the policy of downgrading the rank of taxa.

Formal names of orders and higher categories have not in the past been restricted to rank levels, and there are no ICZN rules to say that they should be treated in accordance with the usage on family and lower levels. Thus, for instance, Broili (1933) coined the name Cheloniellida for an order, while Størmer (1944) used the same name for a subclass, referring to Broili as the author. Herein we follow the general trend and regard Broili as the author for taxa on all levels above the family rank and write 'Subclass Cheloniellida (Broili, 1933) Størmer, 1944' to mark the origin of the term and the later emendation to a new rank level.

In order to sort out possible affinities, we use cladograms to make our data and choices clear (Figs. 87, 88). We see no point in doubling this presentation of the phylogenetic result by introducing it also as the only guideline in the classification, the more so as there is no generally accepted practice of arranging and naming groups in a cladistic system. By contrast, the evolutionary classification is very practical, in addition to conveying important information both on the evolutionary introduction of anatomical novelties and on ancestor–descendant relationships (Charig 1990). We therefore see great informative gains in combining phylogenetic cladograms with evolutionary systematics. We see no problem in defining a so-called paraphyletic group as one that has a monophyletic origin marked by an evolutionary novelty but has not yet acquired key characters of derived groups. This is no more difficult to understand than the definition of a holophyletic group by characters no longer present in many of its members, but it is infinitely easier to handle systematically.

Systematic descriptions

Phylum Schizoramia Bergström, 1976a

Diagnosis. – Arthropods derived from a common ancestor with biramous limbs. (Emended from Bergström 1976a.)

Discussion. – Three well-known groups are included, *viz.* crustaceans, lamellipedians (i.e. trilobitomorphs) and chelicerates. The latter have 'uniramous' limbs, but this is regarded as a modification from the schizoramian condition, and endo- and exopods may be represented in different tagmata. In addition, the Schizoramia include some Cambrian groups, namely *Fuxianhuia*-like forms, *Canadaspis* and allies, *Yohoia* and other megacheirans ('great-appendage arthropods'), *Molaria*-like forms, 'stem-lineage crustaceans', and probably still others.

Superclass Proschizoramia n.supercl.

Name. – Latin *pro*, before, in the sense that the group exhibits characters developed at an early stage in schizoramian evolution.

Diagnosis. – Schizoramian arthropods lacking the apomorphies of derived schizoramian groups, but with group-specific apomorphies, for instance in the tagmosis, development of the 1st post-antennal appendage, and loss of abdominal limbs.

Discussion. – This is considered a 'stem-schizoramian' assemblage of groups.

Class Yunnanata n.cl.

Name. – From Yunnan, the province of China in which the fossils are found.

Diagnosis. – Schizoramian arthropods with multisegmented, stout endopods with unmodified termination, and simple exopod flap without setae; post-antennal legs unmodified except for 1st pair; eye segment separate, followed by rounded portion including ventral mouth and possibly two segments; carapace covering this portion and a few thoracic segments; long body with pleural folds; unstable segmentation or pseudosegmentation with a few pairs of limbs corresponding to each tergite; posterior end with elongate dorsal spine and ventral furca.

Discussion. – Apparent irregularities in the segmentation occurs also in a few other arthropod groups. Thus, in diplopods most body segments are secondarily fused two and two. In euthycarcinoids the pattern of fusion is more irregular. Both of these groups differ from Yunnanata in having uniramous appendages. A more fundamental difference, however, is that the body units carrying two or three pairs of legs are clearly caused by fusion and are two and three times, respectively, as long as single segments. Within Yunnanata, on the other hand, limb pairs are crowded in short segments. Multiplication of limbs occurs in the abdomen of notostracans. In this case, as in Yunnanata, individual body segments carry more than one limb pair. Notostracans are easily distinguished from Yunnanata because they have a whole set of crustacean characters. A separate eye segment in front of the head is not known from other arthropods. The combination of characters in Yunnanata is unique.

Order Fuxianhuiida Bousfield, 1995

Diagnosis. – As for the class.

Family Fuxianhuiidae n.fam.

Diagnosis. – Arthropods of the class Yunnanata with a wide pleural fold in the thorax but none or a very narrow one in the abdomen.

Genus included. – *Fuxianhuia* Hou, 1987.

Genus *Fuxianhuia* Hou, 1987

Type species. – *Fuxianhuia protensa* Hou, 1987

Fuxianhuia protensa Hou, 1987
Figs. 7–11

Synonymy. – □1987b *Fuxianhuia protensa* gen. et. sp.nov. – Hou, pp. 281–282, Pls. 1:1–3; 2:1–4, Text-figs. 1–2. □1991 *Fuxianhuia protensa* – Chen et al., Fig. 1. □1991 *Fuxianhuia protensa* Hou – Hou & Bergström, p. 183, Pl. 2:1. □1991 *Fuxianhuia protensa* Hou – Delle Cave & Simonetta, p. 205, Fig. 8B. □1993 *Fuxianhuia protensa* – Bergström, p. 4. □1995 *Fuxianhuia protensa* Hou – Chen et al. 1995a, pp. 1339–1342, Figs. 1–4.

Holotype. – CN 100126 (Hou 1987b, Pl. 1:1–3) from Maotianshan, level M2.

Other specimens. – CN 100127, CN 110826, CN 115319, CN 115353–115358 and 20 unnumbered specimens.

Fig. 7. Fuxianhuia protensa Hou, 1987. □A, B. CN 115319, from Xiaolantian, level XL1, lower part, after preparation, without filter and with polarized filter, respectively, ×1.5. C, CN 115353, from Maotanshan, level M2, lower part, ×1.8.

Distribution. – Maotianshan, levels M2 and M3; Xiaolantian, level XL1; Meishucum, level K10.

Description. – (a) General characteristics: This animal has the superficial appearance of a merostome. The exoskeleton is well sclerotized on both dorsal and ventral sides. As can be judged from dorsoventral, lateral and inclined views, the body is more or less round in cross section but has ventrolaterally directed pleural folds, which gives dorsoventrally compressed specimens a trilobed appearance. The pleural folds are particularly well developed in the anterior half of the body (Figs. 7, 9A, F, 10A, B, 11B, D). The body ('axial lobe') tapers distinctly forwards in the prothorax. In those specimens where the tergites can be counted to the posterior end, there are 31 tergites plus a telson.

Behind the head, tergite one is only about 28% of the maximum body width (which is in segment ten), tergite three about 60%, tergite four about 72%. In the rear part

of the body the width narrows to around 55% of the maximum width. In the holotype the first six and last 16 tergites are shorter than tergites 7 to 15 (Hou 1987b). This is partially verified in the new material, although it must be remembered that measurements on the fossilized material yield a slightly distorted picture. The following figures therefore should not be read as exact measurements of the living animals. In one specimen (Fig. 11B), the first tergite is about 45%, the third 52%, the sixth 59%, the seventh 81%, and the eighth 100% of the maximal length, acquired in tergites 8 to 12. In another (Figs. 7A–B, 9A) the fourth tergite is about 44%, the sixth 56%, the seventh 63%, and the eighth perhaps 90–100%. There

Fig. 8 (overleaf). *Fuxianhuia protensa* Hou, 1987. □A. CN 115319, detail showing multisegmented appendages, ×7.5. □B. CN 115353, anterior part with antennae and head structures dimly visible, the leg of *Naraoia longicaudata* preserved under the left head shield by chance, ×5.

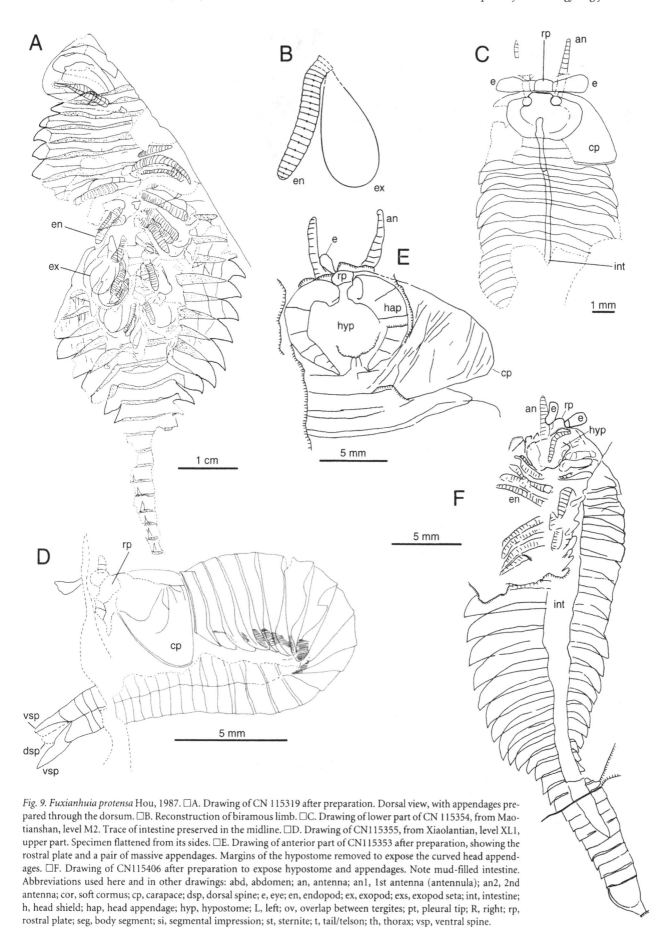

Fig. 9. Fuxianhuia protensa Hou, 1987. □A. Drawing of CN 115319 after preparation. Dorsal view, with appendages prepared through the dorsum. □B. Reconstruction of biramous limb. □C. Drawing of lower part of CN 115354, from Maotianshan, level M2. Trace of intestine preserved in the midline. □D. Drawing of CN115355, from Xiaolantian, level XL1, upper part. Specimen flattened from its sides. □E. Drawing of anterior part of CN115353 after preparation, showing the rostral plate and a pair of massive appendages. Margins of the hypostome removed to expose the curved head appendages. □F. Drawing of CN115406 after preparation to expose hypostome and appendages. Note mud-filled intestine. Abbreviations used here and in other drawings: abd, abdomen; an, antenna; an1, 1st antenna (antennula); an2, 2nd antenna; cor, soft cormus; cp, carapace; dsp, dorsal spine; e, eye; en, endopod; ex, exopod; exs, exopod seta; int, intestine; h, head shield; hap, head appendage; hyp, hypostome; L, left; ov, overlap between tergites; pt, pleural tip; R, right; rp, rostral plate; seg, body segment; si, segmental impression; st, sternite; t, tail/telson; th, thorax; vsp, ventral spine.

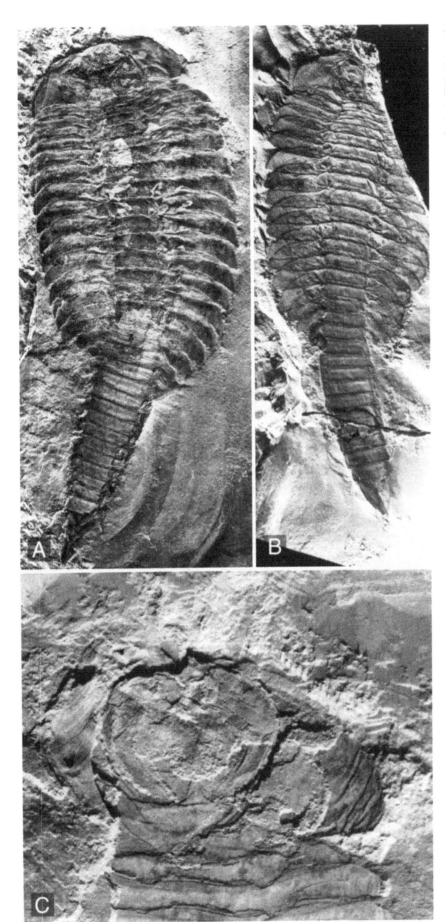

Fig. 10. Fuxianhuia protensa Hou, 1987. □A. CN 110826, from Maotianshan, level M2, lower part, ×1.4. □B. CN115356, from Xiaolantian, level XL1, lower part, ×1.4. □C. CN 115353, anterior part, exposes a pair of massive appendages just behind the antennae and two exopod-like structures on the right side of the head after preparation. Under the left side of the head shield is a leg of *Naraoia longicaudata.* ×5.6.

Fig. 11. Fuxianhuia protensa Hou, 1987. □A. CN 115357 from Maotianshan, level M2, lower part, ×2. □B. CN 115358, from Maotianshan, level M3, lower part, ×1.5. □C, D. CN 115406, from Maotianshan, level M2, upper part, anterior structures shown in ventral view, with mud-filled convex gut, ×4.2. C, panchromatic film, D, orthochromatic film.

is thus a notable change between tergites six and eight, but the details differ a little in each specimen. The step between tergites 15 and 16 is not seen in any of the two specimens just referred to, in which both tergites are some 80–95% of the maximum length. Instead, there is a very gradual and small decrease backwards, so that in the first specimen (Fig. 11B) tergite 23 is still some 80% of the maximum length, and tergite 30 around 75%. This specimen is broken off behind tergite 30.

The smooth curvature in two specimens (Figs. 9D, 11A) demonstrates that the abdomen was flexible and could easily be bent in under the anterior part of the body.

In some specimens an intestine, stuffed with mud, is visible (Figs. 9C, F, 11C, D).

(b) Head: The head has two tergites (Fig. 9C). The anterior one of them is short and may be called a rostral plate. It is bent around the anterior end to cover both the upper and lower sides. A pair of club-shaped structures (Fig. 9C), obviously ending in compound eyes, occurs at each lateral extremity.

The posterior part of the head is much longer. Its tergite has a delicate medial furrow. The underside is described below. A carapace takes its origin between the two parts of the head, i.e. at the posterior margin of the rostral plate. It covers the posterior part of the head and the three-segmented prothorax (Fig. 9C–D). When dorsoventrally flattened, the carapace has an outline close to that of a circle segment.

(c) Body: The body is divided into a trapezoidal anterior part, the prothorax, more or less covered by the carapace, a broad middle part, the opisthothorax, with pronounced pleural folds, and a narrow posterior part, the abdomen, with a triangular tail spine. It gives the animal a remote similarity to a merostome or scorpion.

The prothorax consists of only three tergites, which are notably narrower than the successive ones. The width of the first three tergites is only about 30%, 40%, and 70% of that of the fourth tergite, which in turn is approximately 75% as wide as the tenth tergite (Fig. 11B). The opisthothorax appears to differ slightly in length. Chen *et al.* (1995a) mentioned that the thorax has 17 or occasionally 16 segments. In specimen CN115353 the corresponding number is 18. Thus the opisthothorax in these cases has 13–15 segments. The two anteriormost abdominal segments carry broad pleural folds. Behind these segments the pleural folds are very small (Figs. 7, 9A, D, F, 10A, B). Behind tergites 29–31 there is a keeled, triangular tail spine which roughly equals four tergites in length (Figs. 9D, 10A, B). In side view, the spine is seen to be confined to the dorsal side, and there is a pair of slightly shorter ventral spines (Fig. 9D). The arrangement is thus similar to that of Palaeozoic phyllocarid crustaceans (e.g., Moore & McCormick 1969).

(d) Ventral side: As mentioned above, the anterior part of the head is covered by a rostral plate both dorsally (Fig. 9C) and ventrally (Figs. 9E, 10C). Most of the posterior part of the head is covered by a large hypostome (Figs. 9E, F, 11C. D; the 'anterior ventral plate' according to Chen *et al.* 1995a, who recognized only an anterior fragment of it). It is rounded, with a shallow posterior embayment. On the anterior side there is a pair of antennal embayments, from which the antennae extend forwards. The antennae are short, consisting of about 14 short annuli (Figs. 8B, 9C, F, 10A, 11C, D). Just behind the antennae, and partly concealed by the hypostome, is a pair of wide but fairly flat appendages extending backwards, curving outward and then inward towards the presumed position of the mouth (Figs. 9E, 10C). Each curved appendage consists of probably 8 podomeres, the terminal one being a small blunt cone. The integument appears to be strengthened through thickening in the latter. There is no claw or chela. These appendages are very similar in position and curvature in all specimens, indicating that they had limited mobility. It is likely that much of the movement was performed at the base and in the first articulation, but we see evidence of some mobility also between the distal podomeres. There is evidence of two additional appendages in the head, each with an endopod and a flat exopod.

In the thorax there is a complete pleural doublure, but the area between the pleural bases is not sclerotized. The legs are biramous. The endopod is very sturdy and consists of some 20 short similarly shaped sclerite rings (Figs. 7A–B, 8A, 9A–B). At least on one side of the leg there is a row of tiny articulations between each of the two rings (Figs. 8A, 9B). These articulations are connected by a narrow furrow along the entire endopod. The terminal element is not really a spine, but a swollen cone. There are no spines. The exopod is a broad, oval flap with a reinforced but thin edge; it also appears to have been very thin and flat in life. The margin is entire, without visible spines, setae or hairs. The attachment is not well seen. It cannot be excluded either that the exopod attaches to several endopod segments or that both branches attach separately directly to the body of the animal.

The number of appendage pairs is distinctly larger than the number of body rings. As described above, the anterior 6–7 body rings are notably shorter than the more posterior ones, and it is possible that this corresponds to a difference in the number of leg pairs pro ring, with about two in front and four behind. The middle part of the body, the opisthothorax, appears to have three pairs of legs per body segment except at the posterior end, where there may be up to about four leg pairs for one body segment.

In the abdomen, beginning with one of tergites 17–19, the ventral integument is entirely sclerotized and devoid of openings for appendages. The sclerites are not divided into tergites and sternites but are complete rings. The anterior margin of tergite 19 has a band indicating the onset of this morphology (Fig. 7C). We call this tagma a telosoma, following the nomenclature used for merostomes. The abdomen is protected by complete sclerite rings with rudimentary pleural folds.

Discussion. – Delle Cave & Simonetta (1991, p. 205) illustrated *Fuxianhuia* with 31 instead of 28 abdominal tergites. They considered it related to *Sidneyia*, because what they thought is the head has only three segments. However, the supposed head is the thorax (and, moreover, *Sidneyia* does not have three segments in its head). The two genera are utterly unlike in all other characters.

Fuxianhuia was reinterpreted by Chen *et al.* (1995a), who claimed that they 'reject' the statement (Bergström 1993) that there are more than one pair of appendages per tergite. According to their view, there are two leg pairs corresponding to each segment in the middle part of the thorax and three or possibly four pairs corresponding to the posteriormost tergites. We regard this not as a rejection of our statement, but as support. There are probably three pairs of legs per body segment in most of the thorax, and a larger number (up to about four) in one or a few segments at the posterior end.

Chen *et al.* (1995a) erroneously claimed that Bergström (1993) regarded *Fuxianhuia* as aschelminth- or flatworm-like, apparently because they overlooked the distinction between metamery and pseudomery.

Chen *et al.* (1995a) identified a pair of strong non-chelate appendages in the head of *Fuxianhuia*. We confirm this observation. The presence of these appendages is somewhat surprizing, since the gut in some individuals is densely stuffed with mud (Figs. 9C, F, 11C, D), indicating that *Fuxianhuia* was a mud-eater. However, there the mud commonly contains black grains which may be phosphatic and indicate remains of engulfed prey. We suggest that the appendages were used for shuffling in sediment and prey in the mouth without full discrimination. It should also be mentioned that Chen *et al.* (1995a) use the term *antennule* where we use *antenna*. Whereas *antennule* is not an incorrect term in the context, it is usually applied only in arthropods with two pairs of antennae, i.e. crustaceans.

Bousfield (1995), misinterpreting the description by Chen *et al.* (1995a), regards *Fuxianhuia* as related to euthycarcinids, marrellomorphs and crustaceans.

Family Chengjiangocarididae n. fam.

Diagnosis. – Arthropods of the class Yunnanata with abdomen not differentiated into wide anterior and narrow posterior portions.

Genus included. – *Chengjiangocaris* Hou & Bergström, 1991.

Genus *Chengjiangocaris* Hou & Bergström, 1991

Type species. – *Chengjiangocaris longiformis* Hou & Bergström, 1991.

Chengjiangocaris longiformis Hou & Bergström, 1991

Figs. 12–15.

Synonymy. – □1991 *Chengjiangocaris longiformis* gen. et sp. nov. – Hou & Bergström, 185–186, Pl. 3:5–6. □1991 *Chengjiangocaris* Hou & Bergström – Delle Cave & Simonetta, p. 205, Fig. 8C.

Holotype. – CN 110837 (Hou & Bergström 1991, Pl. 3:5–6; Fig. 12A herein) from Maotianshan, level Cf5.

Other specimens. – One fairly complete specimen exposing appendages, CN 115359, and a few poorly preserved tergites with a series of appendages, CN 115360.

Distribution. – Northwest slope of Maotianshan, level Cf5; Xiaolantian, level XL1; and Fengkoushao village, level FK1, corresponding to M2.

Description. – The holotype (Fig. 12A) consists of five narrow and very short anterior tergites (presumed thorax) and 17 long tergites (abdomen). Head, carapace and tail are missing. The combined length of the five thoracic tergites equals the length of the first abdominal tergite. The abdominal tergites are subequal in length, the posterior ones being only slightly shorter. All tergites extend into pleural folds, which are very narrow in the posteriormost tergites. The first five or so abdominal tergites are widest, the last one only about a third of that width. An indistinct trilobation is seen along the entire body. It may result from the same factors as that in compressed *Fuxianhuia*, in which the pleural folds are bent up to give the impression of trilobation.

CN 115359 is reasonably complete, although the anterior half of the body is poorly preserved (Figs. 12B–C, 14A). This specimen is important in demonstrating that the two other specimens are conspecific, and also in showing the presence of a carapace and the shape of the telson. The legs in the anterior part of the animal are strong but still closely set. Much weaker and very closely set legs are indicated in the anterior part of the body (Figs. 12B, C, 14A; closely set lines also in the upper part of Fig. 14A).

CN 115360 consists of five fragmentary tergites, possibly belonging to the thorax, and remains of 11 successive legs and some unidentifiable fragments (Figs. 13, 14B). The tergites are smooth and small in comparison with the legs. Only one sturdy leg branch is exposed, consisting of up to about 17 short segmental rings and a small conical end piece. A thin ridge extends from one ring to the next. It passes over small nodes at the ring margins; these nodes apparently represent the articulation between neighbouring podomere rings.

Discussion. – Delle Cave & Simonetta (1991, p. 205, Fig. 8C) illustrated *Chengjiangocaris* with five 'head' (i.e. thoracic) and 15 'body' (abdominal) tergites. The separate 'head' tergites led them to speculate on similarities with *Sidneyia*, and because of the number of 'head' segments they thought that it was related to *Sanctacaris* and emeraldellids. In the absence of any known structural similarity between these animals, we see no justification for these suggestions.

The only non-crustacean animal we know of having the same type of thorax is *Fuxianhuia*, which is much more likely to be a relative of *Chengjiangocaris* than any of the other arthropods in the fauna (cf. Fig. 15). Both these two genera have a short carapace covering a short thorax, multisegmented legs with articulation nodes and a simple distal tip without claws, probably more than one pair of legs per body segment, and a conical telson. The multiseg-

Fig. 12. Chengjiangocaris longiformis Hou & Bergström, 1991. □A. Holotype, CN 110837, from Northwest slope of Maotianshan, level Cf5, lower part, ×1.4. □B, C. CN 115359, from Xiaolantian, level XL1, lower part, body bent strongly, ×2.7. B, panchromatic film, C, orthochromatic film.

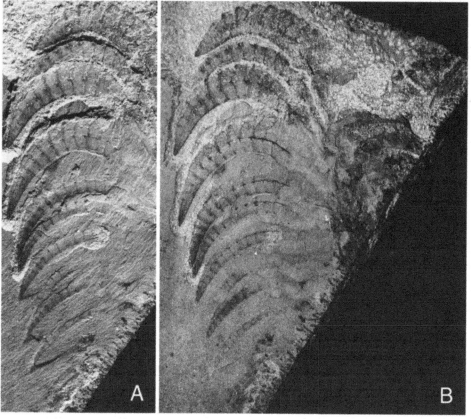

Fig. 13. Chengjiangocaris longiformis Hou & Bergström, 1991. □A, B. CN 115360, from Fengkoushao, level FK1, lower part, panchromatic and orthochromatic films respectively, ×3.1.

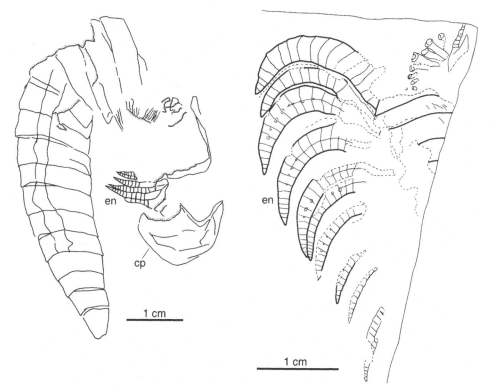

Fig. 15. Chengjiangocaris longiformis Hou & Bergström, 1991. Reconstruction of animal in dorsal view. It is assumed that the head is similar to that of *Fuxianhuia*. For abbreviations, see Fig. 9.

Fig. 14. Chengjiangocaris longiformis Hou & Bergström, 1991. □A. Drawing of specimen CN 115359, lower part, which shows the association of appendages and body, the presence of a carapace, and the tail end. □B. Drawing of CN 115360, lower part, a specimen consisting mainly of endopods. For abbreviations, see Fig. 9.

mentation of the legs and their simple distal tip may be plesiomorphies, as perhaps the presence of more than one pair of legs per segment, whereas the other characters are considered to be synapomorphies. There appear to be no fundamental differences between the two genera. Some thin and overlapping surfaces in *Chengjiangocaris* may be exopods, but they are not well exposed. The difference in spacing between anterior and posterior appendages appears to be more extreme in *Chengjiangocaris* than in *Fuxianhuia*, since the posterior legs are both tiny and very closely set.

Mansuy (1912, p. 31, Pl. 4:6) described a new species from Yunnan under the name of *Amiella prisca*. The type species of *Amiella*, *A. ornata*, was based on a single poorly preserved specimen (Walcott 1911, pp. 27–28, Pl. 5:4). Walcott believed that the genus was closely related to *Sidneyia*, but Hou *et al.* (1995) demonstrated the similarity and possible identity to the anomalocaridid *Peytoia nathorsti* (Walcott 1911). *A. prisca* has definite tergites but no transverse dorsal spine rows and is an arthropod, no anomalocaridid. The presence of fairly large tergites preceded by much shorter ones indicates at least a similarity to *Fuxianhuia* and *Chengjiangocaris*. It may be identical to *C. longiformis*, but the distortion of the specimen makes it difficult to decide with certainty. We regard *A. ornata* as a *nomen dubium*.

Class Paracrustacea n.cl.

Name. – Greek *para*, beside, near, indicating that the animals are not crustaceans, although they show some similarities.

Diagnosis. – Schizoramian arthropods with anteroventral eyes; limbs semipendent; antenna uniramous; 2nd appendage not developed as a 2nd antenna or 'great appendage'; abdomen lacking limbs; carapace gives the animal a crustacean-like habitus; furca-like spines present.

Discussion. – It is important to remember that an animal does not belong to the Crustacea (as generally defined) because of its general habitus or because of a particular number of appendages, but because of very distinctive characteristics such as the presence of a true coxa and a nauplius larva with swimming–feeding specializations, definite signs of which are preserved in the adult. At least one additional limb is involved in the adult set of mouthparts. There is also an advancement, for instance, in the orientation of the exopod (Walossek & Müller 1990).

All these and other characteristics are lacking in arthropods such as *Hymenocaris* and *Canadaspis*. These are superficially more similar to certain advanced crustaceans than to primitive crustaceans, which is one of many examples of convergence among arthropods.

Order Canadaspidida Novozhilov *in* Orlov, 1960

(=Hymenostraca Rolfe, 1969; =Prophyllocarida Simonetta & Delle Cave, 1975)

Discussion. – We agree with Delle Cave & Simonetta (1991, p. 229) that there is no reason to keep the Order Hymenostraca apart from the Canadaspidida. The type species of *Hymenocaris*, *H. vermicauda* Salter, 1853, is poorly known, but the available characters are very close to those of *Canadaspis*: oval narrowing-forwards carapace valves covering the anterior part of the animal, a limbless abdomen with rounded cross section and segments of comparable length and numbers, and characteristically forked and spiny tail appendages (which, admittedly, are longer and more slender in *Hymenocaris* than in *Canadaspis*).

Family Canadaspididae Novozhilov *in* Orlov, 1960

Genus *Canadaspis* Novozhilov *in* Orlov, 1960

Type species. – *Hymenocaris perfecta* Walcott, 1912.

Canadaspis laevigata (Hou & Bergström, 1991)

Figs. 16–21

Synonymy. – □1987c *Perspicaris?* sp. – Hou, p. 297, Pl. 3:6–7. □1991 *Perspicaris? laevigata* sp. nov. – Hou & Bergström, p. 186, Pl. 2:7–8.

Holotype. – CN 110832 (Hou & Bergström 1991, Pl. 2:7). Isolated carapace valve from Maotianshan, level M2.

Other specimens. – CN 100176, open carapace, from Maotianshan, level M2; CN 100178, upper part of isolated carapace valve, from Maotianshan, level Cf4; CN 110833, isolated carapace valve, from Maotianshan, level M3; CN 115361, including lower and upper parts, from Maotianshan, level M2, CN 115362 from Xiaolantian, level XL1, and 25 unnumbered specimens, some of which consist of carapace valves only, while others include soft parts.

Distribution. – Maotianshan, levels M2 and M3, Xiaolantian, level XL1, and Fengkoushao, level FK1.

Description. – Our description is based on all available specimens, not only those illustrated.

(a) General characteristics: Superficially this is a crustacean-like arthropod (Fig. 20) that is basically similar to *Canadaspis perfecta* Walcott from the Burgess Shale, from

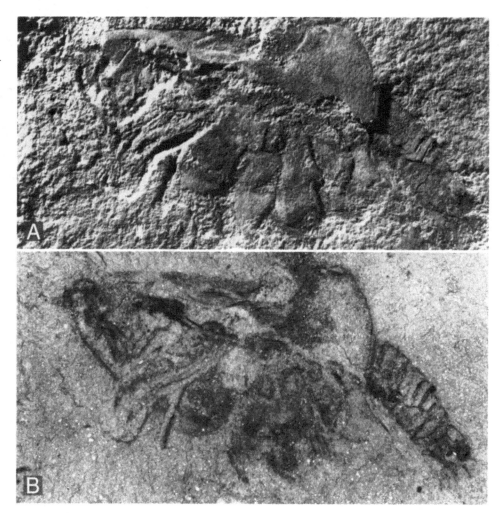

which it differs in having a less spiny telson and a smaller carapace.

(b) Head: The head carries a carapace of very similar shape to that of *C. perfecta*. It is devoid of ornament and distinctly deeper in the posterior part than anteriorly.

(c) Body: All of the body between head and telson are cylindrical and covered by ring-shaped sclerites. These are 19 body tergites, which are somewhat longer in the middle part of the body than anteriorly and posteriorly. The telson is comparatively long; ventrally there is a pair of strong projections from the pre-telson segment, consisting of a strong main spine and a smaller lateral spine.

(d) Ventral side: Anteriorly the head has a pair of stalked eyes and a pair of uniramous antennae, both projecting beyond the edge of the carapace (Figs. 16–20). The rest of the head is difficult to interpret. However, it is clear from the course of the intestinal canal that the head bulges downwards. There is no evidence of any labrum, nor of any specialized mouthparts. There are at least ten pairs of biramous appendages behind the antennae. The first biramous appendage appears to be situated on the head (which is devoid of clear segment boundaries), whereas the succeeding nine belong to the visibly segmented body.

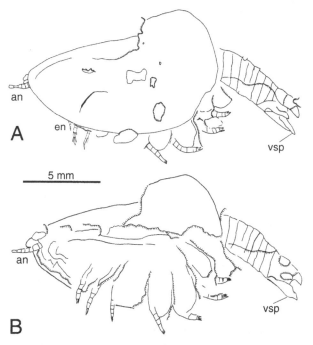

Fig. 17. Canadaspis laevigata (Hou & Bergström, 1991), drawings of CN 115362, lower part, before (A) and after (B) preparation. For abbreviations, see Fig. 9.

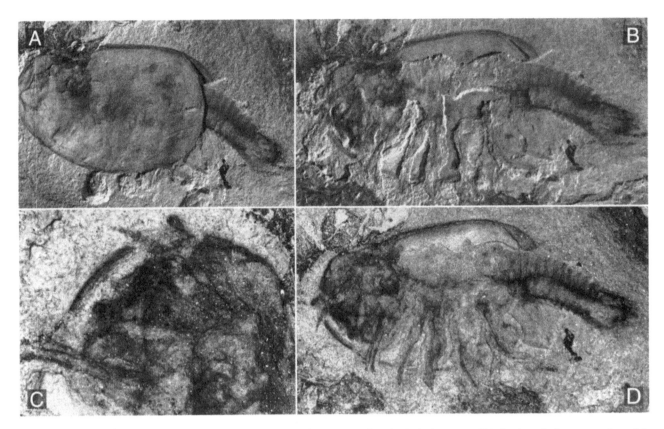

Fig. 18. Canadaspis laevigata (Hou & Bergström, 1991), CN 115361, from Maotianshan, level M2, lower part. □A. Specimen before preparation, ×2.8. □B, D. Specimen after preparation, ×3.1. □C. Enlargement of anterior part, rotated 90°; note eye and antenna, ×7.2. Panchromatic (A, B) and orthochromatic (C, D) film.

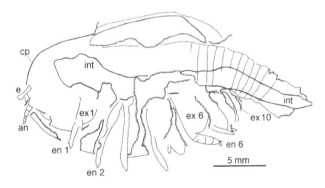

Fig. 19. Canadaspis laevigata (Hou & Bergström, 1991), drawing of CN 115361, after preparation. Note heavy line indicating thick mud-filled gut. For abbreviations, see Fig. 9.

As a result, the abdomen should consist of ten segments, the anterior nine of which are limbless, while the last segment carries tail spines.

All biramous limbs (save that on the pre-telson segment) appear to be of similar shape. The first about four or five limbs are of roughly equal size. Behind, the limbs decrease in size and are more closely set (Figs. 18B, D, 19). The endopod is stout and multisegmented, ending in a set of distal claws alien to crustaceans. The exopod is a flat rounded plate (generally similar to that of *C. perfecta*), the details of which are not yet certain. The attachment areas of the biramous limbs have not been observed. The exopods extend far up the sides of the body, thus indicating a lateral attachment (Fig. 21).

The alimentary canal extends forwards and upwards from the mouth, and then backwards straight to the end of the telson (Figs. 16, 17B, 18B, D, 19). In some specimens it is filled with mud and quite thick (Figs. 18B, D, 19).

Discussion. – The new species is sufficiently similar to the Canadian *C. perfecta* that a close relationship is clear.

Briggs (1978b, Fig. 29) reconstructed the appendages as extending from the ventral side of the body. However, in both our and Briggs's (e.g., 1978b, Figs. 50, 58, 81, 84) material, the exopods extend high on the lateral sides of the body. One possible explanation is that the exopods expand dorsally from the proximal part. An exopod illustrated by Briggs (1978b, Figs. 95, 97, 103) does not show any such expansion, but it may come from the posterior end of the thorax. An alternative explanation is that the limb attached high on the side, but this does not appear likely unless there was quite a long attachment.

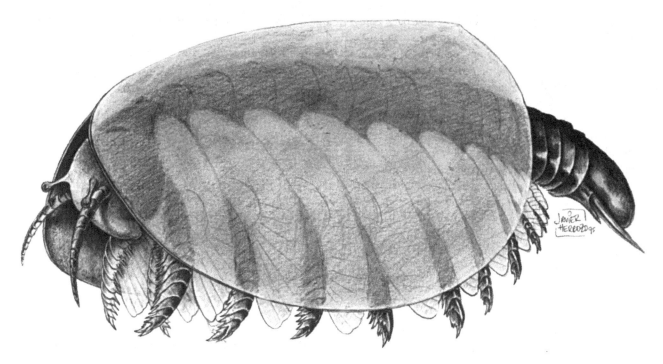

Fig. 20 (above). *Canadaspis laevigata* (Hou & Bergström, 1991), reconstruction of animal in anterolateral view. The mouth presumably could be extended downwards to ingest mud from the substrate.

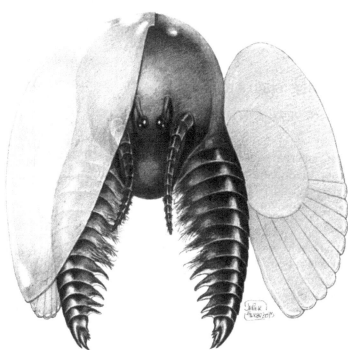

Fig. 21 (right). *Canadaspis laevigata* (Hou & Bergström, 1991), reconstruction of animal in frontal view, with eyes, antennae and one pair of biramous appendages exposed. One half of the bivalved carapace is drawn to show how the exopod must have been flexed back under it. To the right the exopod is folded out.

Briggs (e.g., 1978b, 1992) originally thought that *Canadaspis perfecta* is a malacostracan crustacean; we accept that they appear to have about the same number of segments. The Chinese species, however, has a larger number of segments. Whereas malacostracans have 19 segments between the eyes and the telson, *C. laevigata* has at least 21 (including one pair of antennae and one or two pairs of legs in the head plus 19 in the body). This invalidates any identification of *Canadaspis* tagmata with those of malacostracans.

Briggs regarded the protruding uniramous antennae of *Canadaspis* as 2nd antennae. This would be odd for a malacostracan, since the uniramous state of the 1st antenna and biramous (in cases triramous) state of the 2nd antenna are characteristic features of crustaceans. The 2nd appendage (the third in Briggs' original view) was interpreted by Briggs as a mandible, although he noted that the two in the pair are widely separated (Briggs 1992, p. 297 and Fig. 10). The appendages of the head, as well as those of the body, are attached high up on the sides

(Figs. 18B, D, 19), and only the tips would have been able to reach the mouth opening. There is no possibility that a basal ('coxal') endite could have approached the mouth area. Following Manton (1978), the appendage could therefore be compared to a whole-limb jaw (although it is no jaw at all from a functional point of view), but not to a 'coxal' jaw. The problem with Briggs' interpretation is that whole-limb jaws are present in myriapods and insects, while crustaceans have 'coxal' jaws.

The lack of specialization of the three anterior appendage pairs implies that *Canadaspis* lacked a nauplius larva, a distinctive crustacean characteristic. We note that the leg structure in *Canadaspis laevigata* is identical to that of *C. perfecta*. The spiny termination of the endopods and the peculiar exopods, devoid of setae, also are features alien to crustaceans.

The mouth of *Canadaspis* is fairly far back on the head, probably behind the structure interpreted by Briggs (1992, Fig. 10) as a hypostome which may be the somewhat keel-shaped ventral side of the head in front of the mouth. This ventral position of the mouth, the lack of functional jaws and other specialized limbs, and the mud-filled gut all indicate that *Canadaspis* was a sluggish sediment-eater.

Class Megacheira n.cl.

Name. – Greek *mega*, large, great, and *cheir*, hand.

Diagnosis. – Schizoramian arthropods with eyes, when present, anteriorly placed; limbs pendent; 1st antenna commonly reduced, 2nd limb developed as a 2nd antenna (or 'great appendage'); pleural fold present; elongate terminal tergite, no furca.

Discussion. – These are the 'great appendage arthropods' of, e.g., Bergström (1992), in which the enlarged anterior appendage appears to be the frontalmost appendage. For the first time, a uniramous antenna positioned in front of the 'great appendage' is now described from a member of this group, *Fortiforceps foliosa*. The 'great appendage' thus corresponds to the 2nd antenna of crustaceans. This conclusion is in accord with the 2nd antenna in most arthropods being a slender, multiarticulated, uniramous appendage.

Among the arthropods of the Chengjiang fauna, *Leanchoilia* is placed in the Order Leanchoiliida Størmer, 1944, and *Jianfengia* in the Yohoiida Simonetta & Delle Cave, 1975. The new species *Fortiforceps foliosa* belongs in the same general grouping but should perhaps be placed in a different order.

We have not described *Jianfengia* herein, since we do not have significant new information. *Jianfengia* is fairly similar to *Leanchoilia*, but the frontal appendage is more closely comparable with that of *Yohoia*. The most obvious difference between *Jianfengia* and *Yohoia* is that the former has biramous legs throughout the tail, whereas the latter has a few posterior segments without legs. The latter situation is obviously derived and is no argument against affinity.

Order Leanchoiliida Størmer, 1944

(nom. corr. Størmer 1959, ex. Leanchoilida Størmer, 1944)

Family Leanchoiliidae Raymond, 1935

Genus *Leanchoilia* Walcott, 1912

Type species. – *Leanchoilia superlata* Walcott, 1912.

Leanchoilia illecebrosa (Hou, 1987)

Figs. 22–30

Synonymy. – □1987a *Alalcomenaeus? illecebrosus* sp.nov. – Hou, p. 253, Pls. 3:1–3; 4:1–2. □1991 *Alalcomenaeus? illecebrosus* Hou – Hou & Bergström, p. 183, Pl. 3:1–3. □1991 *Alalcomenaeus illecebrosus* Hou – Delle Cave & Simonetta, p. 218.

Holotype. – CN 100124, from the northwest slope of Maotianshan, level Cf2.

Other specimens. – CN 100125, CN 110834–110836, CN 115363–115371. *Leanchoilia illecebrosa* is one of the most common arthropods in the Chengjiang fauna; several hundred specimens with soft parts have been available for the description.

Distribution. – Maotianshan, primarily level M2, but also M3; Northwest slope of Maotianshan, levels Cf2, Cf3, Cf5, Cf6; Xiaoliantan, level XL1; Fenkoushou, level FK1; Jianbaobaoshan, level Dj1; and Ma'anshan, level Ma1.

Description. – (a) General characteristics: The animal is somewhat shrimp-like in its general habitus, although it lacks a carapace fold. It is notably shorter than *Fortiforceps*, having only 11 (as compared with 20) segments between head-shield and telson. This makes it easy to distinguish these genera even without appendage characteristics.

Most specimens are laterally compressed, indicating that the animal was not dorsoventrally flattened in life. A mud-filled gut is occasionally seen (Figs. 26, 27C, 28B).

(b) Head: Head length is about the length of 2.5–3 body segments. Head flattens notably forwards and ends in a short spine, in the same way as the Burgess Shale *Leanchoilia superlata* (Fig. 30). The head shield appears smooth, devoid of sessile eyes.

(c) Body: The body consists of 11 segments of virtually identical length (Fig. 23). Each tergite appears to form a

Fig. 22. Leanchoilia illecebrosa (Hou, 1987), CN 115363, from Xiaolantian, level XL1, lower part. □A. Lateral view of entire animal, ×4.9. □B. Anterior portion, ×10.8. □C. Posterior portion, ×16.5.

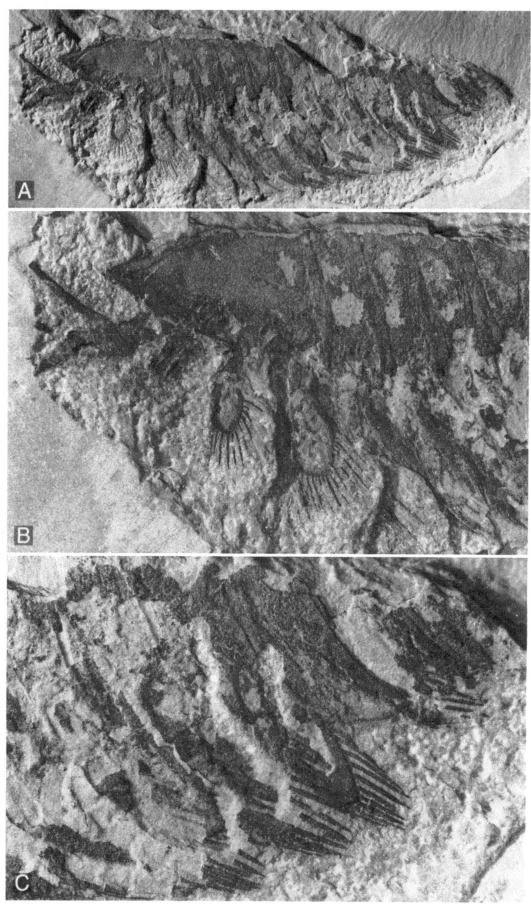

Fig. 23. Leanchoilia illecebrosa (Hou, 1987), CN 110834, from Maotianshan, level M2, lower part, entire animal in lateral view, after preparation further (cf. Hou & Bergström, 1991, p. 184, Fig. 1), ×5.1. Panchromatic (A) and orthrochromatic (B) film.

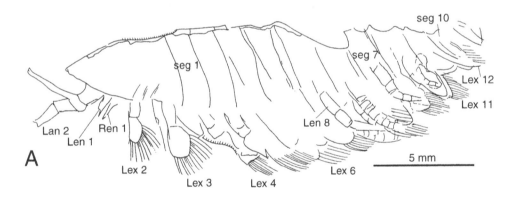

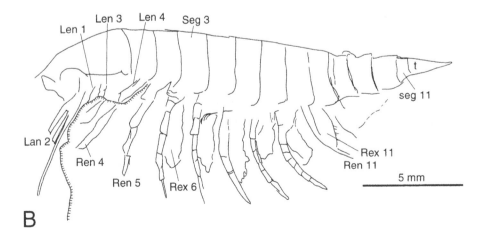

Fig. 24. Leanchoilia illecebrosa (Hou, 1987). ☐A. Drawing of CN 115363, lower part. ☐B. Drawing of CN 110834, lower part. For abbreviations, see Fig. 9.

Fig. 25. Leanchoilia illecebrosa (Hou, 1987), CN 115364, from Xiaolantian, level XL1, lower part, lateral view, ×10.4. Panchromatic (A) and orthochromatic (B) film.

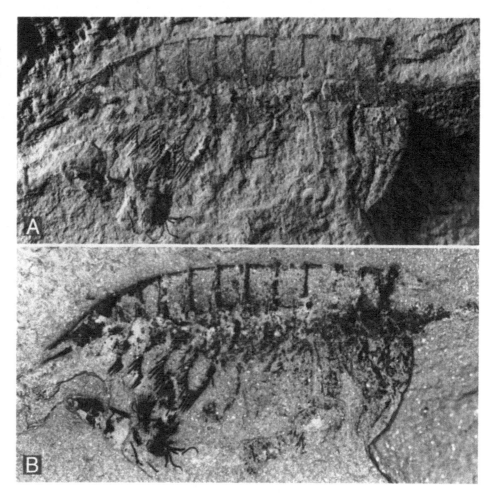

Fig. 26. Leanchoilia illecebrosa (Hou, 1987), CN 115365, from Maotianshan, level M3, lower part. Anterior part of animal in lateral view, ×13. Panchromatic (A) and orthochromatic (B) film.

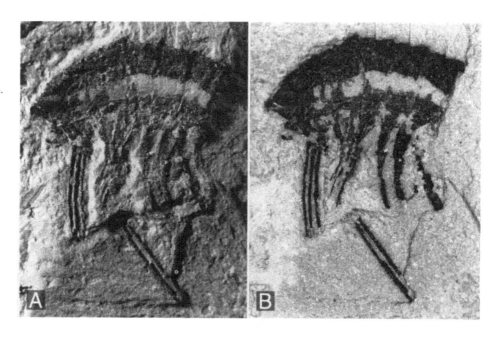

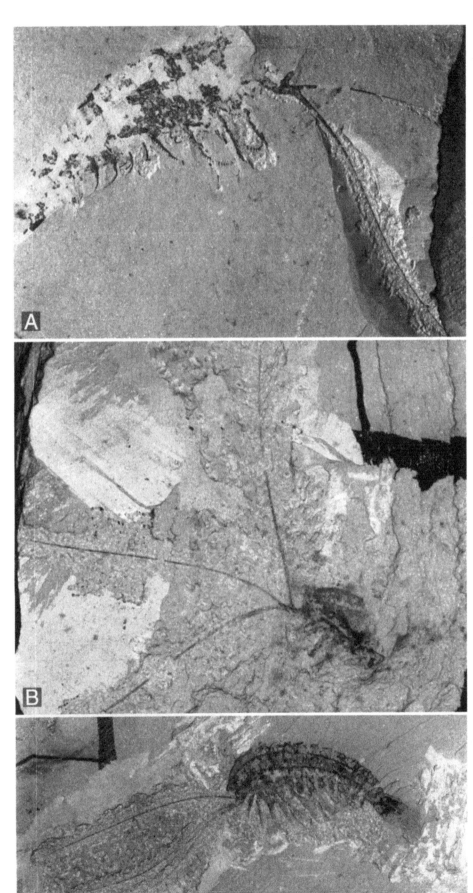

Fig. 27 (this page). *Leanchoilia illecebrosa* (Hou, 1987), with large 2nd antenna. ☐A. CN 115366, from Ma'anshan, level Ma1, upper part, ×4.5. ☐B. CN 110835, from Maotianshan, level M2, upper part, after further preparation (cf. Hou & Bergström, 1991, p. 184, Fig. 2), ×4.5. ☐C. CN 115367, from Jianbaobaoshan, level Dj1, lower part, ×2.7.

Fig. 28 (opposite page). *Leanchoilia illecebrosa* (Hou, 1987). ☐A. CN115368, from Jianbaobaoshan, level Dj1, lower part, posterior part with tail in lateral view, ×22.4. ☐B. CN 115369, from northwest slope of Maotianshan, level Cf5, lower part, specimen with tail in dorsal view and with exposed appendages in anterior end, ×4.4. ☐C. Anterior end of CN 115369, exposing appendages with endo- and exopod, ×14.

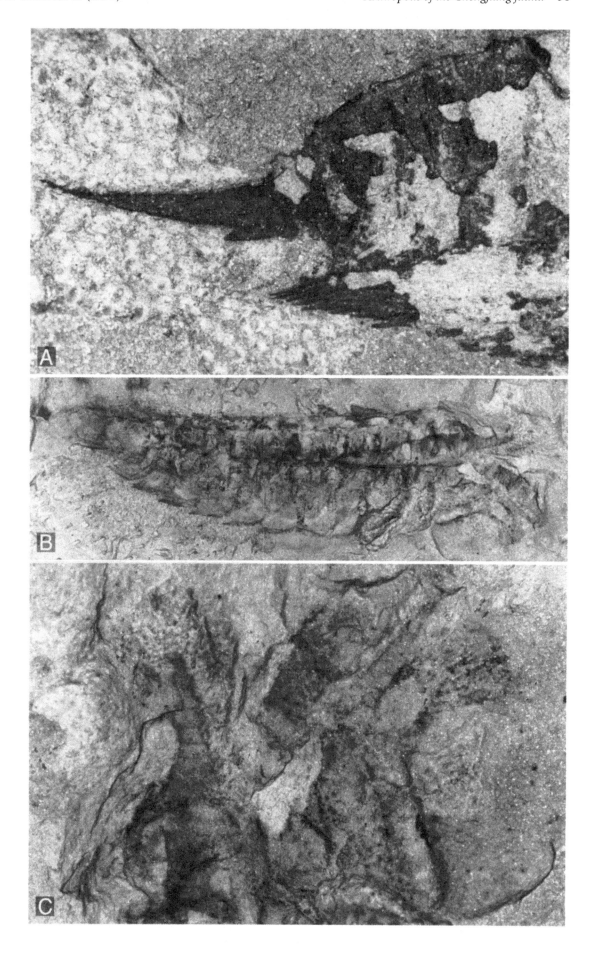

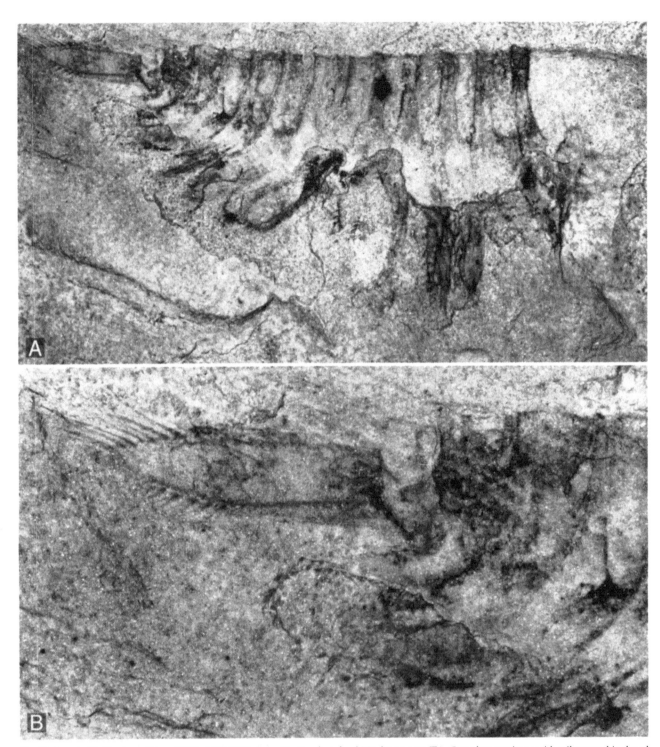

Fig. 29. Leanchoilia illecebrosa (Hou, 1987), CN 115370, from Maotianshan, level M2, lower part. □A. Complete specimen with tail exposed in dorsal view, ×5. □B. Posterior part showing marginal spines, ×13.

smooth semi-cylinder. The anterior half of the body is of about equal width and height; the posterior part tapers slightly.

The telson has a wide and high anterior part, is articulated to the last segment, and is posteriorly extended into a flat surface, which is widest behind the middle and ends in a terminal acute angle (Fig. 29). At least anteriorly the

flat surface is surrounded by sloping flanges. In the posterior part, the edges carry strong, tapering spines (Fig. 29).

(d) Ventral side: In a dorsoventrally flattened specimen (CN 115371), a pair of black structures close to the anterior border possibly represent paired ventral eyes (Fig. 30). Several specimens confirm that these structures occur in front of the 2nd antennae. In lateral view the

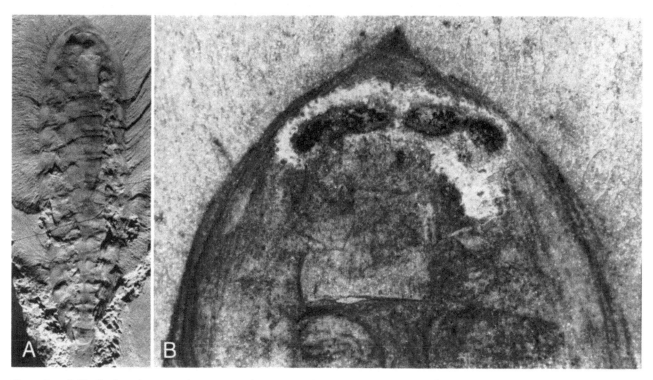

Fig. 30. Leanchoilia illecebrosa (Hou, 1987), CN 115371, from Maotianshan, level M2, lower part, animal in dorsal view. □A. Entire animal, ×1.8. □B. Anterior end to show possible eyes, ×13.

structure appears as a thick shaft separated from the rounded eye by a constriction.

Two morphological types of appendages are seen, the 2nd antennae and the succeeding biramous appendages. There is no trace of a 1st antenna. The 2nd antenna has a fairly short and robust proximal portion (Fig. 22A–B), beyond which two bifurcations result in three narrow distal branches, each of which is very long (about as long as the body from tip to tail end) (Fig. 27). The segments of the branches are quite long, very much longer than the annuli of typical (1st) antennae.

The head carries three additional pairs of appendages, similar to those of the body in being divided into endopod and exopod, but smaller (Figs. 22A, B, 24A).

The exopods are clearly exposed in different specimens (Figs. 22, 25, 29). The distal element is a rounded flap, which carries strong tapering setae or spines, like those of the telson. The spines are usually sub-equal in size, but occasionally there is an apparent alternation between strong and weak spines (Fig. 22B). It is possible that these represent two different rows situated on the same margin, each row consisting of about ten spines; if so, the setae of the two rows may be of equal size, but those of the row beneath appear thinner because of poor exposure.

Endopods are commonly seen (Figs. 22, 23, 24A–B, 29A) but are not often well preserved. However, one specimen clearly reveals endopods as well as the contact between exo- and endopod (Fig. 28B–C).

Discussion. – The originally poorly known species was questionably placed in the poorly known genus *Alalcomenaeus* (Hou 1987a). Delle Cave & Simonetta (1991, p. 218) realized that it is closely comparable with the Burgess Shale *Leanchoilia superlata* (see Gould 1989, Figs. 3.52–54). The dorsal exoskeleton of *L. illecebrosa* may be somewhat narrower, but the head is of similar shape (including the anterior prolongation into a blunt spine); the number of trunk segments is identical, and the tail in both species has marginal spines. Both also lack signs of a 1st antenna, and in both the 2nd antenna is extended into three very long flagellae. The distal part of the exopod is virtually identical in the two species; the rest of the biramous appendage is not known in *L. superlata*.

Order Fortiforcipida n. ord.

Diagnosis. – Megacheiran arthropods with short head, long undifferentiated body and a telson bearing five broad plates; eye stalked; 1st antennae present; 2nd antennae robust, with fingers as in *Leanchoilia*, but without flagellae and succeeded by undifferentiated biramous legs (three pairs on head and one pair corresponding to each thoracic tergite); exopod simple, consisting of a plate with radiating pattern and short spines surrounding outer margin; endopod long and multisegmented, quadrate in cross section.

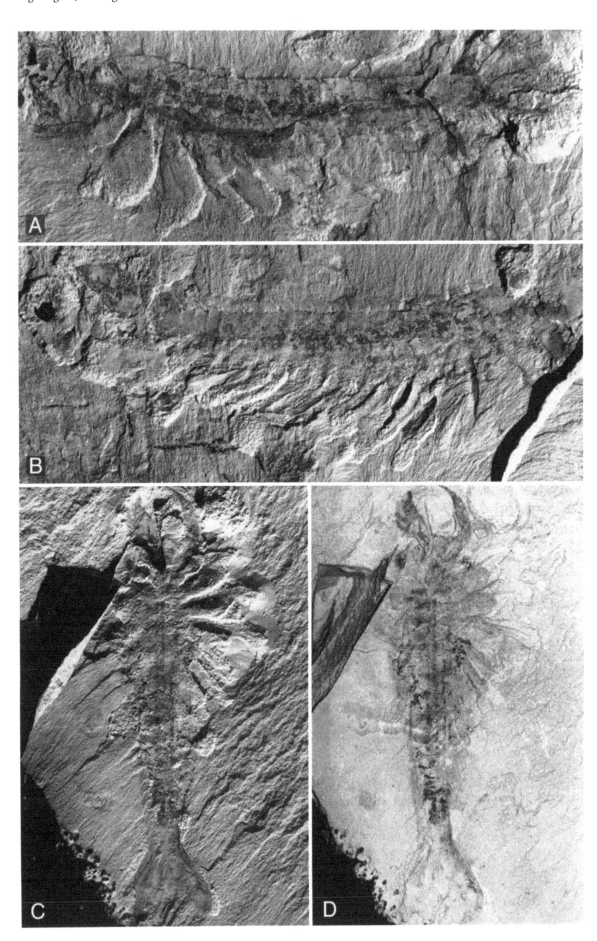

Family Fortiforcipidae n. fam.

Diagnosis. – As for the order.

Genus included. – *Fortiforceps* n. gen.

Genus *Fortiforceps* n. gen.

Name. – Latin *fortis*, robust, and Latin *forceps*, pincers, referring to the 2nd antennae.

Type species. – *Fortiforceps foliosa* n. gen. et sp.

Diagnosis. – As for the family.

Fortiforceps foliosa n. gen. et sp.
Figs. 31–35

Name. – Latin *foliosus*, full of leaves, referring to the foliate tail.

Holotype. – A complete, obliquely compressed specimen with well-preserved appendages, lower and upper parts, CN 115372 (Fig. 31A–B), from level M2 at Maotianshan.

Other specimens. – Four additional specimens. CN 115373, from level M2 at Maotianshan, lower and upper parts, a complete specimen showing an obliquely dorso-ventral compression. CN 115374 and 115375, both from level XL1 at Xiaolantian and having lower and upper parts, two lateral specimens with fine endopods. One poorly preserved specimen, lower part only (not illustrated).

Distribution. – Maotianshan, level M2, and Xiaolantian, level XL1.

Description.. – (a) General characteristics: Elongate, shrimp-like animal with a general body shape reminiscent of a narrow truncated cone. The maximum width is at the third tergite, from which the body tapers progressively backwards to the tail segment. The two complete specimens are 3.9 cm (Fig. 31A–B) and 3.7 cm (Fig. 31C–D) long, excluding the 2nd antennae. The body consists of 20 segments and a telson with five wide plates constituting the tail (Figs. 31A, C, D, 33D, E). The head bears stalked eyes, paired 1st and 2nd antennae, and three pairs of biramous legs (Figs. 31B, 32, 33). Each tergite of the body

Fig. 32. *Fortiforceps foliosa* n. gen. et sp. ☐A. CN 115374, from Xiaolantian, level XL1, lower part, showing granulose ornaments and spines on the 2nd antenna and approximately quadrate endopods in cross section, ×3.8. ☐B, C. CN 115375, from Xiaolantian, level XL1; B, lower part, ×2.3; C, upper part, ×4.6.

corresponds to a pair of biramous legs. The appendages are more or less three-dimensionally preserved. The biramous legs of the head show the same characteristics as, but are smaller than, those of the body (Figs. 31A, C, D; 33D, E).

The head and body are laterally (Figs. 32, 33A), obliquely (Figs. 31A–B, 33C, D) or dorsoventrally (Figs. 31C, D, 33E) compressed. The broad tail tends not to be laterally compressed. The exoskeleton and appendages are purple, while the intestine is black. These parts stand out well against the yellowish matrix. The intestine is straight and simple, extending from the middle of the

Fig. 31 (opposite page). *Fortiforceps foliosa* n. gen. et sp. ☐A, B. Holotype, CN 115372, from Maotianshan, level M2, upper and lower parts, respectively, oblique compression, with pleural tips of the right side visible on top, ×3.6. ☐C, D. CN 115373, from Maotianshan, level M2, lower part, oblique dorsoventral compression, panchromatic and orthrochromatic films, respectively, ×2.7.

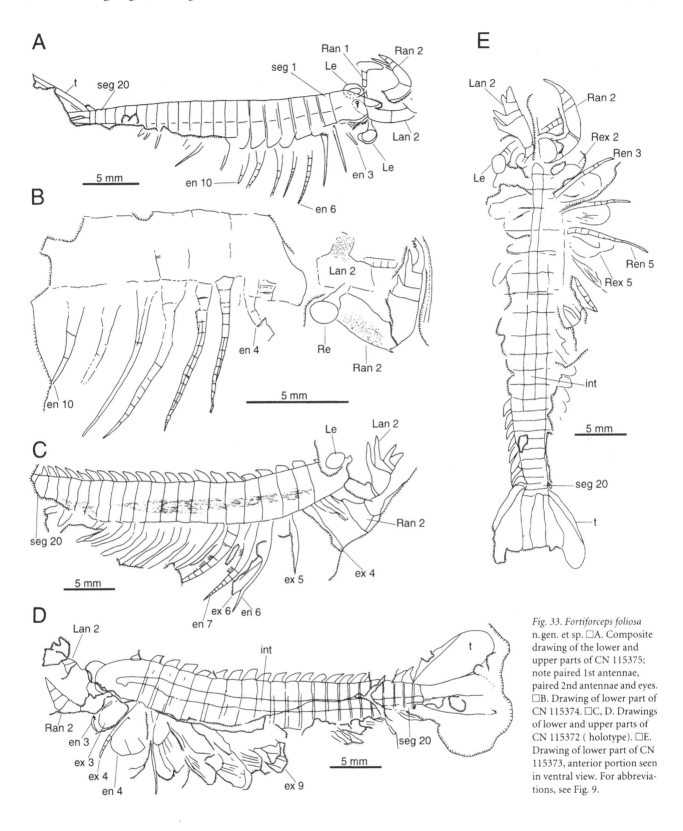

Fig. 33. Fortiforceps foliosa n.gen. et sp. □A. Composite drawing of the lower and upper parts of CN 115375; note paired 1st antennae, paired 2nd antennae and eyes. □B. Drawing of lower part of CN 115374. □C, D. Drawings of lower and upper parts of CN 115372 (holotype). □E. Drawing of lower part of CN 115373, anterior portion seen in ventral view. For abbreviations, see Fig. 9.

head to the anus at the end of the telson; it does not show any filling (Figs. 31, 33C–E).

(b) Head: The head shield is best visible in the laterally and obliquely compressed specimens. The anterior margin is seen in one of them, partly through a compound eye

(Fig. 31A, 33D). In the other complete specimen the anterior margin is concealed by the 1st and 2nd antennae as well as by the eyes (Figs. 32B–C, 33A). It is possible, however, that the margin is still visible as an oblique, slightly curved line. In a dorsoventrally compressed specimen,

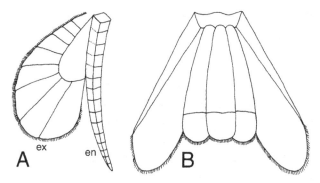

Fig. 34. Fortiforceps foliosa n.gen. et sp. □A. Reconstruction of biramous limb. □B. Reconstruction of tail in dorsal view. For abbreviations, see Fig. 9.

preparation has partially exposed an evenly curved anterior margin (Figs. 31C–D, 33E). The more often exposed posterior margin is transverse.

(c) Body: The body is covered by 20 trunk tergites and comprises also a tail with five elongate flaps. In the laterally and obliquely compressed specimens, the body is broadest at the third trunk tergite and gradually narrows backwards (Figs. 31A, 32B, 33A, C, D). The 2nd and 3rd tergites are also the longest in the series. Each segment boundary is shown by two lines, indicating the amount of overlap between successive tergites. The tergites are produced laterally into short pleural spines, which are well exposed on the upper side of the obliquely compressed specimen (Figs. 31A–B, 33C, D). The pleurae are more light-coloured than the rest of the exoskeleton.

The tail has an anterior unit articulated to the preceding tergite (Figs. 31C–D, 33E, 34B). It gives rise posteriorly to three lanceolate plates, each with the distal tip set off by an articulation, and posterolaterally to a pair of larger lateral plates. All five plates are bordered posteriorly by bristles.

(d) Ventral side: Anteriorly there is a pair of large globular stalked eyes (Figs. 31–33, 35). No details of the eye surface have been observed. The stalk, or peduncle, is long

and seemingly segmented. The eyes face various directions, indicating flexibility of the eye stalks.

A possible 1st antenna is seen in some specimens (Figs. 31A–B, 32, 33A–B, E). In one specimen the antenna ends in a distal swelling (Figs. 32B, 33A), but it is uncertain whether this is a typical condition or a malformation. The 1st antenna attaches near the anterolateral margin of the head.

Behind the eyes, but still in the anterior part of the head, are the 2nd antennae, the site of attachment of which is well seen (Figs. 31C–D, 32B, 33A–E). In ventral view, a large oval impression shows the opening of the proximal segment of the 2nd antenna (Fig. 33E). This antenna consists of six segments. The proximal segment is short, segments 2–3 longer, 4–6 again short. Segment 6 is the terminal element and is a simple cone, although there appears to be some variation in the detailed morphology (cf. Figs. 33B–C and E). Segments 3–5 each carries a prominent immobile projection on its medial side. The 2nd antenna is covered with small pustules (Figs. 32A, 33B).

All appendages behind the 2nd antennae are biramous and of the same general design. The endopod has about 15 podomeres; the proximal one or two are shorter than wide, the others are slightly longer than wide. Each podomere is approximately quadrate in cross section (Figs. 32A, 33B). The exopod is similar to that of *Canadaspis perfecta* (See Briggs 1978) and *C. laevigata*; an oval flap consisting of an inner lobe and an outer rim split up by a number of weak radiating lines into about 8 sectors (Figs. 31A, 33D). The outline of the inner lobe is much the same as that of the entire exopod. The outer margin is set with fine setae or bristles.

Discussion. – Fortiforceps is herein assigned to megacheiran arthropods but, alternatively, it may represent a new group of proschizoramians. The 2nd antennae are most similar to those of *Yohoia* and *Jianfengia*, having strong distal spines. As in *Jianfengia*, there are biramous appendages all the way to the tail. *Fortiforceps* is unique in apparently having well-developed 1st antennae in addition to

Fig. 35. Fortiforceps foliosa n.gen. et sp. Reconstruction in lateral view.

the 'great appendages'. This is of considerable significance, since it indicates that the 'great appendage'of the other forms is not a 1st antenna, but the 2nd appendage, a 2nd antenna in some way comparable to that of crustaceans. The exopods of *Fortiforceps* are unique among megacheiran arthropods in having radial lineations (as in *Canadaspis*) and in lacking strong needle-like setae. The tail is a fan that can be compared only with tail fans in some genuine crustaceans.

We have reconstructed the antenna with a peculiar distal club (Fig. 35), although it is seen in only one specimen (see above).

Fortiforceps appears to occupy a fairly isolated systematic position among the megacheiran arthropods. The presence of appendages on all of its segments supposedly is a primitive feature.

Proschizoramia?

Class uncertain

Order Acanthomeridiida n. ord.

Name. – From the family name Acanthomeridiidae n. fam.

Diagnosis. – Arthropods with well-developed pleura; tergum not mineralized; head shield with free cheeks separated by suture, body with eleven tergites and a narrow tail spine.

Families. – Acanthomeridiidae n. fam.

Family Acanthomeridiidae n. fam.

Name. – From the generic name *Acanthomeridion* Hou, Chen & Lu, 1989.

Diagnosis. – As for the order.

Genus included. – *Acanthomeridion* Hou, Chen & Lu, 1989.

Genus *Acanthomeridion* Hou, Chen & Lu, 1989

Type species. – *Acanthomeridion serratum* Hou, Chen & Lu, 1989

Diagnosis. – As for the family.

Acanthomeridion serratum Hou, Chen & Lu, 1989

Fig. 36

Synonymy. – ☐1989 *Acanthomeridion serratum* gen. et sp. nov. – Hou *et al.* pp. 55–56, Pls. 3:1–5; 4:1–5; Text-figs. 3–4. ☐1991 *Acanthomeridion/Acanthomerion* Hou, Chen & Lu – Delle Cave & Simonetta, p. 218, Fig. 17D.

Holotype. – CN 108305 (Hou *et al.* 1989, Pl. 4:1–2) from Maotianshan, level M2.

Other specimens. – CN 108306–108310 (Hou *et al.* 1989) and seven not illustrated specimens, one of which is preserved with a tail spine.

Distribution. – Maotianshan, levels M2, M3, M4, Cf3.

Description. – (a) General characteristics: The body is divided into a head and a body with eleven broad tergites and a narrow tail spine. The back appears to have been evenly vaulted. The tergites are extended into pleural spines. All specimens are dorsoventrally compressed, which indicates that the animal in life was somewhat flattened and did not enroll. The appendages are unknown.

(b) Head: The lateral outline of the head is almost parabolic. The posterior margin appears to be straight. The genal angles are extended into very short, pointed genal spines. The surface is smooth. There are no eyes, and the only notable feature of the head is a pair of small free cheeks separated from the main part of the shield by sutures.

(c) Body: The body is almost parallel-sided, tapering only slightly from the first segment to the tenth. The posterior margins of the tergites are serrated (hence the species name). The tergites have pleura with short spines in the anterior segments, but longer spines posteriorly. Tergite 9 has a particularly long and slender spine; tergite 10 is less extended. The anterior tergites appear fairly straight, while the posterior ones are strongly arched forwards. This arching varies and is very little pronounced in specimen CN108306 (Hou *et al.* 1989, Pl. 3:3–4). Tergite 11 arches back to embrace a thin tail spine separated by a basal articulation (Fig. 36).

Discussion. – The variation in the arching of the posterior trunk segments indicates that the arching is very much an effect of tilting forwards before compression. This in turn gives an indication of the original convexity. The posterior part of the trunk was apparently broadly hogbacked, with flat sides sloping ventrolaterally. The axial part appears to have been much wider anteriorly and the pleura narrower, which means that this part of the body probably had a much more rounded vaulting.

The rectangular outline is thus only an effect of the difference in the width of the pleural lobes, which dimin-

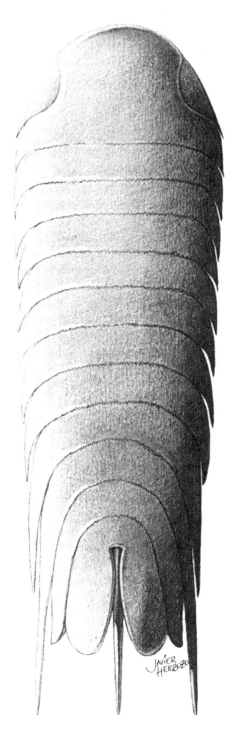

Fig. 36. Acanthomeridion serratum Hou *et al.*, 1989. Reconstruction of animal in dorsal view. New features are free cheeks and terminal spine.

ishes the similarity of *A. serratum* to myriapods and other uniramians. *Acanthomeridion* is therefore in all probability a schizoramian arthropod, but without knowledge of its appendages it cannot be placed systematically with certainty.

Superclass Crustaceomorpha Chernyshev, 1960

Emended diagnosis. – Schizoramian arthropods, mostly with pendent limbs, slender, articulated exopods with needle-shaped setae, and flattened furcal rami (or secondarily modified).

Discussion. – This group comprises the 'pan-crustaceans' of German authors (see Walossek & Müller 1990), including 'stem-lineage crustaceans' and true crustaceans, i.e. the classes Pseudocrustacea and Crustacea.

Class Pseudocrustacea Størmer, 1944

Emended diagnosis. – Crustaceomorphs without the full set of crustacean characters; in particular, nauplius larva, labrum(?) and coxal segment not developed, the latter being represented by a 'proximal endite'; furca originally having flattened rami.

Discussion. – Størmer (1944, pp. 134–135) introduced the Class Pseudocrustacea as a subdivision of the subphylum Trilobitomorpha. Three orders were included, *viz.* (in Størmer's spelling) the Burgessida, Waptida, and Hymenocarina. None of these appears to belong with the trilobitomorphs (lamellipedians). The position of burgessiids is uncertain. The forms included by Størmer in the Hymenocarina are now distributed in the orders Hymenostraca, Canadaspidida, Protocaridida and Odaraiida, all of which are of crustacean-like rather than trilobite-like habitus. Waptiids also have a crustacean-like habitus, and appear to belong to the crustaceomorphs. Størmer selected no type subgroup of the Pseudocrustacea; we select the Waptiida for this purpose.

The group is taken to include the 'stem-lineage crustaceans' of Walossek & Müller (1990) as well as other arthropods that appear to be on the same branch of the schizoramian tree but have not yet evolved all crustacean characters. Thus, the larva has more than three pairs of legs and is not a nauplius larva, and there is a corresponding lack of distinction of three pairs of specialized anterior limbs in the adult. Furthermore, there is no coxa and no labrum, and the exopod setae commonly are directed away from the endopod rather than towards it.

We think that *Canadaspis*, one of the best known crustacean-like forms from the Burgess Shale, should be interpreted in the light of these features. Briggs (1978b) demonstrated its superficially malacostracan-like habitus and placed it among the phyllocarid crustaceans. However, Dahl (1987) did not accept it as a true crustacean, and particularly not as a malacostracan. If we try to apply Walossek & Müller's (1990) criteria, firstly it lacks the typical crustacean larva with three pairs of appendages.

We note that the presence of such a larva is revealed in the adult of primitive Upper Cambrian *orsten* crustaceans such as *Hesslandona*, *Walossekia*, *Skara*, *Bredocaris* and *Rehbachiella*, in the shape of a large, multisegmented and setose exopod in the 2nd antenna and mandible, while the successive limbs are quite different. In contrast, in *Canadaspis* the supposed mandible is serially similar to the successive thoracic legs and the supposed 2nd antenna (regarded as probably the 1st antenna by Dahl 1984) is uniramous and devoid of setae. Regarding setation, not only are there no exopod setae directed towards the endopod, as in early true crustaceans, but there are not even setae on the other side, as in the 'stem-lineage crustaceans' of Walossek & Müller (1990). There is no indication of any coxa, nor of a labrum. It is clear that *Canadaspis* is not a crustacean in the sense of Walossek & Müller (1990) and, moreover, that it cannot even be very close to the true crustaceans. The near-correspondence in tagmosis between *Canadaspis* and 'higher' crustaceans must be a case of convergent evolution. Bivalved carapaces appear to have evolved many times; for example, even small ostracode-like crustaceans have evolved more than once, as demonstrated in the ostracodes, phosphatocopids and bradoriids (Hou *et al.* 1996).

What, then, is the correct systematic position of *Canadaspis*? Similarities with crustaceans not discussed above are the possession of carapace and a supposed furca. A carapace occurs widely outside the true Crustacea (e.g., in *Fuxianhuia*), and the supposed furca is very different from what we see in crustaceans. It is similar only to the corresponding structure in *Hymenocaris*. Briggs (1978b, p. 482) claimed that these spines in *Canadaspis* and *Hymenocaris* represent different structures, the former in the last segment, the latter in the telson. We find it difficult to see such a difference. The two genera are also very similar to each other in side view of the carapace and in the shape and number of segments in the abdomen. *Hymenocaris* is said to lack a carapace hinge. The significance of this difference should not be exaggerated; the lack of a hinge is an original and larval character, and the hinge might disappear in the adult by paedomorphosis.

The absence of exopod setae and the development of the exopod as a rounded flap makes *Canadaspis* comparable to *Fuxianhuia* and to the megacheiran arthropods.

Order Waptiida Størmer, 1944

(nom. corr. Størmer 1959, ex. Waptida Størmer, 1944)

Family Waptiidae Walcott, 1912

(nom. corr. Størmer 1959, ex. Waptidae Walcott, 1912)

Genera included. – *Waptia* Walcott, 1912, and *Chuandianella* Hou & Bergström, 1991, are tentatively included in the family.

Genus *Chuandianella* Hou & Bergström, 1991

Type species. – *Mononotella ovata* Li, 1975.

Chuandianella ovata (Li, 1975)

Fig. 37

Synonymy. – □1975 *Mononotella ovata* sp. nov. – Li, p. 65, Pl. 3:16–17. □1975 *Mononotella viviosa* sp. nov. – Li, p. 66, Pl. 3:18. □1975 *Mononotella marginia* sp. nov. – Li, p. 66, Pl. 3:19–20. □1982 *Mononotella viviosa* Li – Jiang, p. 214, Pl. 29:10, 12. □1982 *Mononotella subquadrata* sp. nov. – Jiang, p. 214, Pl. 29:14. □1985 *Mononotella ovata* Li – Huo & Shu, p. 167, Pl. 29:1–2. □1985 *Mononotella marginia* Li – Huo & Shu, p. 167, Pl. 29:3–4. □1985 *Mononotella viviosa* Li – Huo & Shu, p. 167, Pl. 29:5–6. □1985 *Mononotella subquadrata* Jiang – Huo & Shu, p. 167, Pl. 29:7. □1985 *Mononotella longa* sp. nov. – Huo & Shu, p. 168, Pl. 29:8. □1985 *Mononotella chuanshaanensis* sp. nov. – Huo & Shu, p. 168, Pl. 29:9–10. □1985 – *Mononotella alta* sp. nov. – Huo & Shu, p. 169, Pl. 30:1–3. □1985 *Mononotella dianshaanensis* sp. nov. – Huo & Shu, p. 169, Pl. 3:4–5. □1991 *Mononotella ovata* Li – Huo *et al.*, p. 182, Pl. 40:1–2. □1991 *Mononotella marginia* Li – Huo *et al.*, p. 183, Pl. 40:3–4. □1991 *Mononotella viviosa* Li – Huo *et al.*, p. 183, Pl. 40:5–6. □1991 *Mononotella subquadrata* Jiang – Huo *et al.* p. 183, Pl. 40:7. □1991 *Mononotella longa* Huo & Shu – Huo *et al.* p. 183, Pl. 40:8. □1991 *Mononotella chuanshaanesis* Huo & Shu – Huo *et al.* p. 184, Pl. 40:9–10. □1991 *Mononotella alta* Huo & Shu – Huo *et al.* p. 184, Pl. 41:1–3. □1991 *Mononotella dianshaanensis* Huo & Shu – Huo *et al.* p. 185, Pl. 41:4–5. □1991 *Chuandianella ovata* (Li) – Hou & Bergström p. 186, Pl. 2:5–6.

Holotype. – Y010, carapace only (Li, 1975, Pl. 3:16), from Qiongzhusi in Kunming, level YN 6303. The holotype is housed in Chengdu Institute of Geology and Mineral Resources, Chengdu, Sichuan Province, China.

Other specimens. – CN 110830 and 110831, carapace only; CN 115376 with soft parts, from Maotianshan, level M2, seven unnumbered specimens with soft parts and some hundreds of carapaces.

Distribution. – Southern Shaanxi, Sichuan, Yunnan and Guizhou Provinces, Eoredlichia Zone, Lower Cambrian Qiongshusi Formation.

Discussion. – *Chuandianella ovata* has not yet been studied in detail. The general design (Fig. 37) is similar to that of *Waptia* and *Plenocaris* and very different from *Canadaspis* and *Perspicaris*. For example, it is distinguished from *Canadaspis* by having a rectangular carapace rather than a pear-shaped, by long antennae and body segments, and by the bilobed tail.

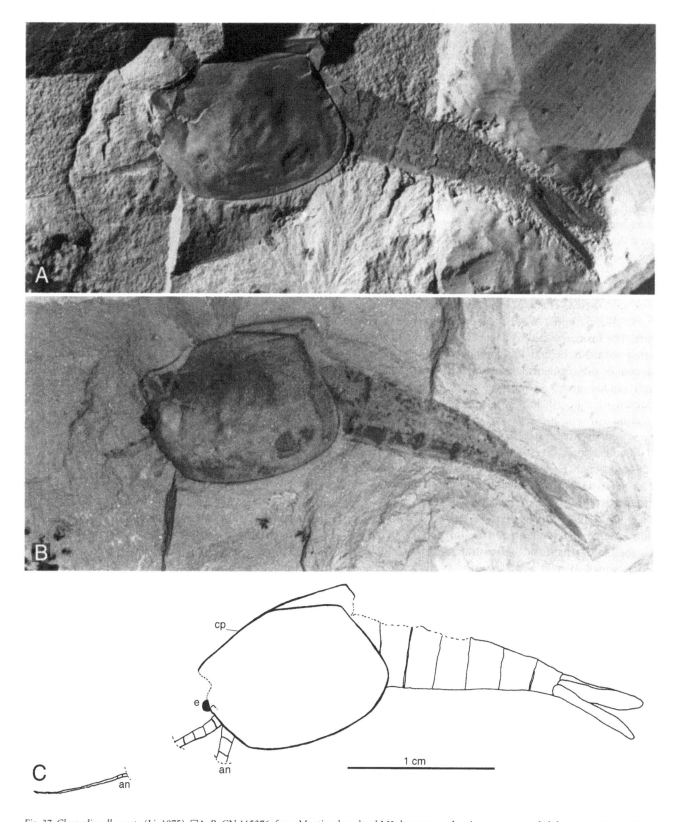

Fig. 37. Chuandianella ovata (Li, 1975). □A, B. CN 115376, from Maotianshan, level M2, lower part, showing antennae and abdomen, ×4.5. Panchromatic (A) and orthrochromatic (B) film. □C. Drawing of CN 115376. Note long antennae. Scale 1 cm. For abbreviations, see Fig. 9.

Chuandianella ovata was originally described on the large valves alone, but we have now identified the body. On the whole, it conforms with *Waptia fieldensis* and *Plenocaris plena* from the Burgess Shale. *Plenocaris* was described by Whittington (1974). *Waptia* has been discussed by several authors but has not so far been redescribed within the project on the Burgess Shale fauna initiated by H.B. Whittington. Because of the imperfect knowledge of all three genera, it is difficult to evaluate the relationships between them. Whittington (1974, pp. 19–20) noted as a difference between *Waptia fieldensis* and *Plenocaris plena* that the caudal furca is segmented in the former but not in the latter. *Waptia fieldensis* appears to have a slenderer body than the other two species, while *Chuandianella ovata* has notably long antennae and a relatively larger carapace than the others. Tagmosis has produced particularly short thoracic and long abdominal segments in *Waptia*. There is also a distinct difference between *Waptia* and *Plenocaris* in the number of segments. The former is said to have 13 or 14 appendage-bearing segments behind the antennae and a limbless abdomen of five segments (see below), the latter a limbless(?) head, a thorax of four segments and a limbless abdomen of nine (including telson). The differences would place *Waptia* and *Plenocaris* in different subclasses if they were extant crustaceans. However, they are not, and differences must be estimated differently in the Cambrian evolutionary bush. In *Chuandianella* the situation is not yet known, and the affiliation with the Waptiidae is only tentative.

Tiegs & Manton (1958, pp. 292, 314), referring to Heldt, stressed the similarity between *Waptia* and the protozoea larva of peneaeid malacostracans. Briggs (1983, p. 5) regarded *Waptia* and *Plenocaris* as most closely related to *Canadaspis*, although at the same time he regarded *Plenocaris* as probably a non-crustacean, *Waptia* as a possible relative of the Branchiopoda (Briggs 1983, p. 6), and *Canadaspis* as a representative of an advanced crustacean group, the malacostracan Phyllocarida (Briggs 1978b, 1992). His different conclusions are mutually incompatible. Whittington (1977) regarded *Plenocaris* as a member of the Phyllocarida.

It has often been said that *Waptia* looks like a crustacean. There is a carapace-like structure, as in many crustaceans. Anteriorly the head has a pair of compound eyes and a pair of uniramous antennae. What follows behind the antennae is not quite clear. According to Briggs (1983, pp. 5–6), there may be 14 (13?) paired appendages behind the (1st) antenna. These should include a reduced 2nd antenna, three (two?) small (uniramous) appendages, four (presumably uniramous) walking legs, and six limbs provided with a 'gill-branch'. These limbs belong to fairly short segments. More posteriorly, there is a limbless abdomen with five comparatively long segments. The body is terminated by a telson with flat furcal blades.

The evidence for uniramous limbs behind the antenna in *Waptia* is weak. For instance, in the 7th limb from the posterior, there is apparently a well-developed outer branch, albeit somewhat shorter than in the successive limb (specimen U.S.N.M. 57682, cf. Walcott 1912, Pl. 27:5, and Simonetta & Delle Cave 1975, Pl. 41:1A–B). On the other hand, many specimens expose a difference in habitus between the anterior and posterior limb series. The inner branch clearly dominates anteriorly, the outer branch posteriorly. The outer branch in all probability is an exopod. Its setae are directed inwards, towards the endopod. This is a morphology that originated in the 'stem-lineage crustaceans' (Walossek & Müller 1990). The flat furcal blades appeared roughly at the same evolutionary place. However, the absence of a large 2nd antenna is strong evidence that *Waptia* is not a true crustacean (Walossek & Müller 1990).

Class Crustacea Pennant, 1777

Subclass Branchiopoda Latreille, 1817

Order Odaraiida Simonetta & Delle Cave, 1975

Family Odaraiidae Simonetta & Delle Cave, 1975

The Chengjiang fauna contains no undoubted crustacean. A possible exception is a specimen of *Odaraia?* sp. (Fig. 38), which exhibits a long terminal element typical of many branchiopods.

Superclass Lamellipedia n. supercl.

(Subphylum Arachnomorpha Heider, 1913, emend. Størmer, 1944 [as phylum]; Subphylum Trilobitomorpha Størmer, 1944)

Name. – Latin *lamella*, diminutive of *lamina*, plate, and *pes* (*pedi-*), foot.

Diagnosis. – Arthropods with extended pleura (in marrelomorphs only in the head region), a semipendent stance or lateral deflection of the whole limb and an exopod 'comb' with flattened setae as apomorphic characters. The compound eyes are originally ventral, but there is a strong tendency to shift them to the dorsal side of the head.

Discussion. – The reality of Størmer's group Trilobitomorpha has often been questioned. However, it is quite clear that a group of trilobites and trilobite-like arthropods are held together by a small but distinctive and consistent set of characters (Bergström 1992; cf. diagnosis). The arthropods from the Chengjiang fauna add much

Fig. 38. Odaraia? eurypetala Hou & Sun 1988. CN 115377, from Mao-tianshan, level M2, lower part. Posterior part of animal. The short abdominal segments are characteristic of calmanostracan branchio-pods, ×4.

information to this debate and considerably strengthen the concept of the Lamellipedia. The trilobites have generally been referred to a distinct class, although Manton (1978) regarded them as a phylum, while Whittington (1977, 1985a) went the opposite way and included some other lamellipedians in the Class Trilobita. We agree with Whittington that the group should not form a phylum or subphylum. On the other hand, we believe that the name Trilobita is so closely associated with the true trilobites with calcified tergum that it would be most unfortunate to use this name for all the lamellipedians. As the term Trilobitomorpha was tied to the phylum level by Størmer, we prefer another name for the class level taxon.

There are some class-level names used for the (non-trilobite) lamellipedians: Subclass Aglaspidida Bergström, 1968; Order Cheloniellida Broili, 1933, used as subclass by

Størmer (1944); Subclass Emeraldellida Størmer, 1944; Order Marrellomorpha Beurlen, 1934, used as class by Størmer (1944); Class Merostomoidea Størmer, 1944; Subclass Prochelicerata Størmer, 1944; Subclass Pseudo-notostraca Størmer, 1959; Class Trilobitoidea Størmer, 1959; Subclass Xenopoda Raymond, 1920. In addition, Starobogatov (1985) launched a series of new names and emended others. None of all these names appears suitable as the name of a superclass embracing the trilobito-morphs. This is the reason to introduce such a name here, a name that also focuses on the diagnostic morphology of the exopod setae.

The lamellipedians can be divided into a number of groups. We follow Størmer (1944) in distinguishing a group that appears to have been the first to branch off as the Class Marrellomorpha. The others are united herein to form the new class Artiopoda, with a number of sub-classes (see below).

Class Artiopoda n. cl.

Name. – Greek *artios*, complete, even, and Greek *pous* (*podo-*), foot, referring to the complete set of similar appendages behind the antennae.

Diagnosis. – Lamellipedians arthropods with trilobite-like appearance, broad tergum, and usually a complete set of fairly undifferentiated post-antennal appendages.

Subclasses. – Nectopleura n. subcl., Conciliterga n. subcl., Trilobita Walch, 1771, Petalopleura n. subcl., Xenopoda Raymond, 1935, and Aglaspidida Bergström, 1968.

Subclass Nectopleura n. subcl.

Name. – Greek *nektes*, swimmer, and Greek *pleura*, side, referring to the name of the Order Nectaspidida and the wide pleural fold surrounding the body.

Diagnosis. – Artiopodans with large tail shield and semi-pendent limbs with a wrinkled proximal cormus; compound eyes in original ventral position; feeding through mud ingestion.

Orders. – Nectaspidida Raymond, 1920, and Retifaciida n. ord.

Order Nectaspidida Raymond, 1920

(nom. corr. herein, ex Nectaspida Raymond, 1920 [=Naraoidea Størmer, 1944; =Naraoiformes Starobogatov, 1985])

Families. – Liwiidae Dzik & Lendzion, 1988 (p. 35, not formally defined), with *Liwia* Dzik & Lendzion, 1988 (replaces homonym *Livia* Lendzion, 1975), *Tariccoia*

Hammann, Laske & Pillola, 1990, and *Soomaspis* Fortey & Theron, 1995; Naraoiidae with *Naraoia* and *Maritimella* Repina & Okuneva, 1969; Orientellidae Repina & Okuneva, 1969 with *Orientella* Repina & Okuneva, 1969.

Discussion. – This group is characterized by a large tail tending to comprise the entire body behind the head. Unlike the Helmetiida, which are characterized by a similar fusion, adjoining pleura overlap along the entire back, and there are no rostral or pararostral plates. The Liwiidae (*Liwia*, *Tariccoia* and *Soomaspis*) have head and tail shields and 3–4 thoracic tergites; the Naraoiidae (*Naraoia* and possibly *Maritimella*) have only head and tail shields; and the Orientellidae (*Orientella*) have probably head and tail shields in addition to one short and one very long thoracic tergite. We agree with Dzik & Lendzion (1988) that, based on *Liwia convexa* (Lendzion, 1975), *Liwia* is most probably a close relative of *Naraoia*, and we therefore place the Naraoiidae and Liwiidae in the same order. Fortey & Theron (1995) regard the Liwiinae to be a subfamily of the Naraoiidae, which is reasonable.

Family Naraoiidae Walcott, 1912

Emended diagnosis. – Nectaspidids with tergum divided into large head and tail shields, with no intervening thoracic tergites.

Discussion. – The family was recently discussed by Fortey & Theron (1995), who stressed the importance of heterochronic processes in the evolution of naraoiids.

Genus included. – *Naraoia* Walcott, 1912.

Genus *Naraoia* Walcott, 1912

Type species. – *Naraoia compacta* Walcott, 1912.

Discussion. – In phylogenetic analyses, *Naraoia* has been assigned to either the trilobites, because of the number of its cephalic appendages and the development of a tail shield (Briggs & Whittington 1981; Briggs 1983, 1990; Fortey & Theron 1995), or to the non-trilobite lamellipedians (trilobitomorphs; e.g., Størmer 1944; Ramsköld & Edgecombe 1991; Bergström 1992).

Naraoia longicaudata Zhang & Hou, 1985

Figs. 39–45

Synonymy. – ☐1985 *Naraoia longicaudata* sp.nov. – Zhang & Hou, p. 594, Pls. 1:1–2; 2:2–4; 3:1–4. ☐1991 *Naraoia* – Chen *et al.*, Fig. 6. ☐1991 *Naraoia longicaudata* Zhang & Hou – Delle Cave & Simonetta, p. 199, Fig. 5G. ☐1994 ?*Naraoia* sp.nov. – Erdtmann *et al.*, Fig. 6. ☐1996

Naraoia longicaudata Zhang & Hou – Ramsköld *et al.* 1996, pp. 16–18, Fig. 2. ☐1996 *Naraoia longicaudata* Zhang & Hou – Ramsköld & Edgecombe, pp. 271–273, Fig. 1E.

Holotype. – CN 94354, a lower part (Zhang & Hou 1985, Pl. 1:1) from Maotianshan, level M2.

Other specimens. – CN 94355, 94357–94363 (Zhang & Hou 1985), 115315–115317, 115378–115384, several hundred specimens with appendages, and thousands of articulated and disarticulated shields.

Distribution. – Maotianshan, levels M2, M3 and M4; northwest slope of Maotianshan, levels Cf1–Cf8; Xiaolantian, levels XL1 and XL2; Fengkoushou, level FK1, Jianbaobaoshan, levels Dj1, Dj2; Dapotou, Ma'anshan and Hongjiashong, horizon corresponding to M2, M3.

Naraoia spinosa Zhang & Hou, 1985

Figs. 46–47

Synonymy. – ☐1985 *Naraoia spinosa* sp.nov. – Zhang & Hou, p. 594, Pls. 2:1; 3:5; 4:1–3. ☐1991 *Naraoia spinosa* Zhang & Hou – Delle Cave & Simonetta, p. 199, Fig. 5C. ☐1991 *Naraoia?* sp. – Hou *et al.*, p. 404, Fig. 5.

Holotype. – CN 94365, lower part (Zhang & Hou 1985, Pl. 4:1) from Maotianshan, level M2.

Other specimens. – CN 94356, 94364–94367, 115281–115284, 115385–115387 and several hundred specimens with soft parts and thousands of articulated and disarticulated shields.

Distribution. – Maotianshan, levels M2 and M3; northwest slope of Maotianshan, levels Cf1–Cf6; Xiaolantian, levels XL1, XL2; Fengkoushao, level FK1; Jianbaobaoshan, levels Dj1, Dj2; Dapotou, Ma'anshan and Hongjiachong, horizon corresponding to M2, M3.

Preliminary account of the two NARAOIA *species.* – Because of the large quantity of material of the two *Naraoia* species from the Chengjiang fauna, a separate description of each is planned. Thus we present only a summary of preliminary observations and conclusions rather than formal descriptions and discussions.

Fig. 39. Naraoia longicaudata Zhang & Hou, 1985. ☐A. CN 115378, from Maotianshan, level M2, lower part, exposing a long antenna and demonstrating the relief in the exopods of the right side, ×1.9. ☐B. CN 115379, from Northwest slope of Maotianshan, level Cf5, lower part, after preparation, ×1.9; see also Fig. 40A. ☐C. CN 115380, from Maotianshan, level M2, upper part, with nicely exposed endopods and exopods, ×1.9; cf. Fig. 40B. ☐D. CN 115381, from Maotianshan, level M2, lower part, showing mudstuffed gut, ×1.9; cf. Fig. 40C.

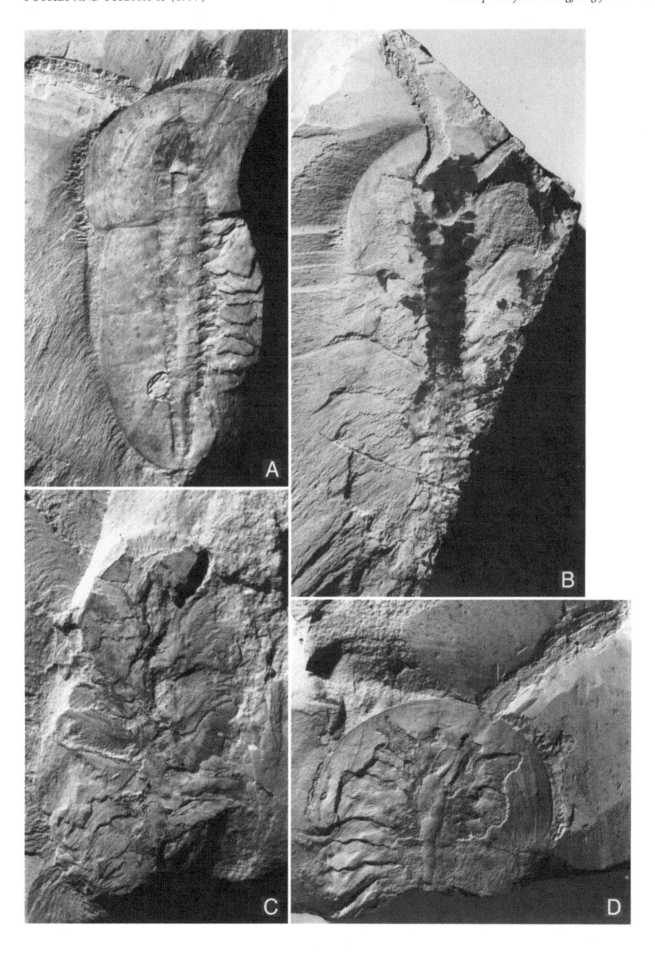

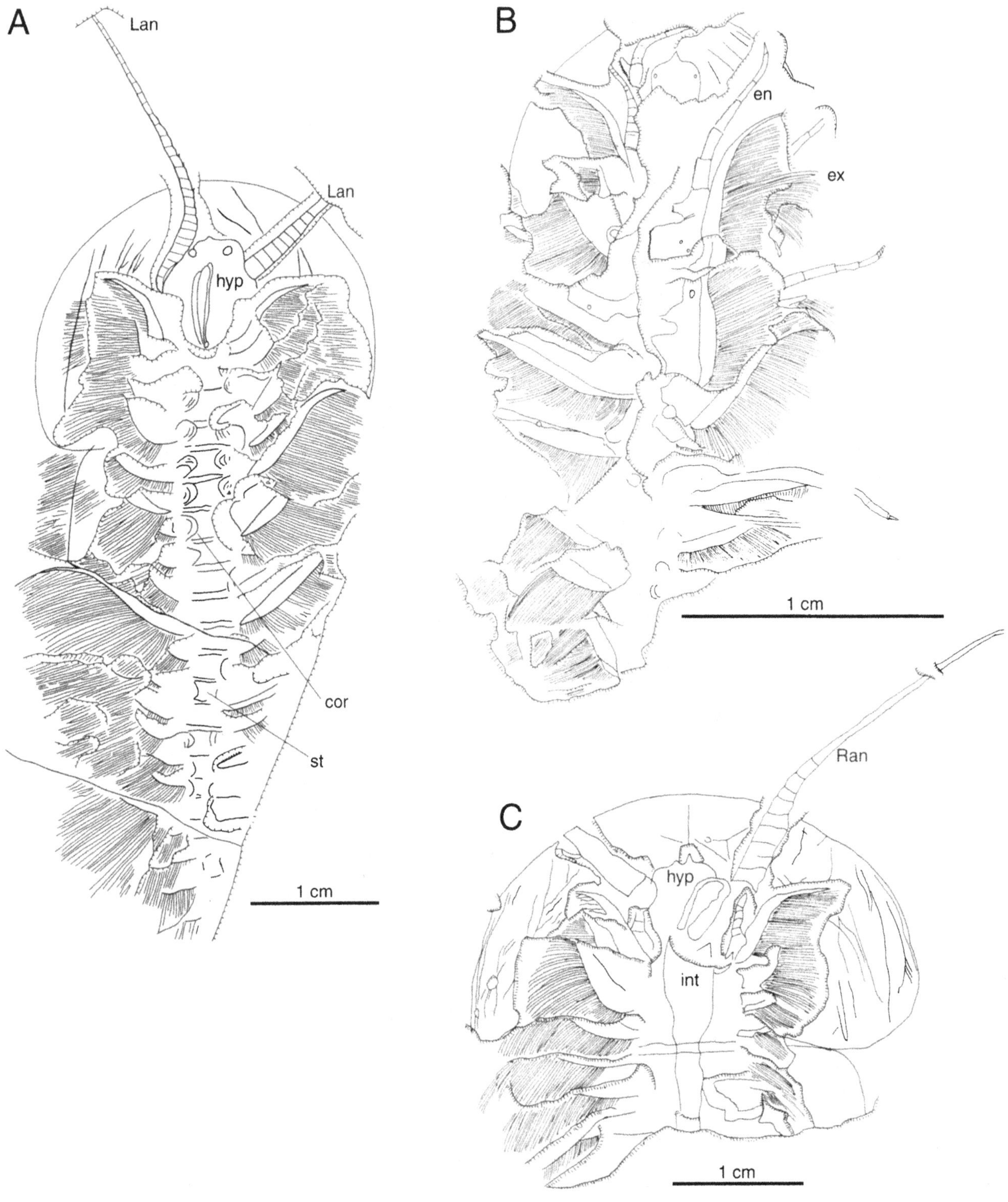

Fig. 40. Naraoia longicaudata Zhang & Hou, 1985. □A. Drawing of CN 115379, in dorsal view after preparation, attaching position of antennae, anterior intestine and position of mouth, probable eyes on anterolateral hypostome and ringed cormus compressed on the surface of exoskeleton. □B. Drawing of CN 115380, split along appendages, showing flattened-out limbs on the left which appear to have a 45° anti-clockwise rotation. □C. Drawing of CN 115381, in dorsal view after preparation, relief intestine. For abbreviations, see Fig. 9.

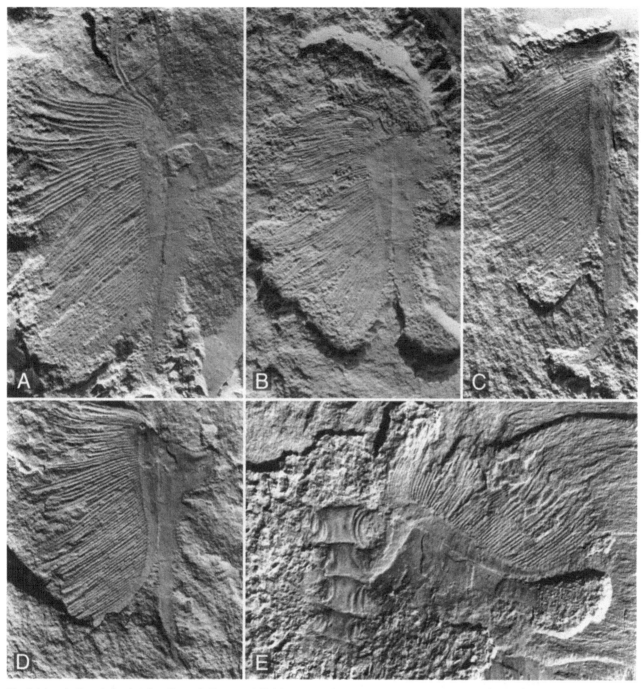

Fig 41 (above). *Naraoia longicaudata* Zhang & Hou, 1985. □A–D. Isolated appendages showing exopod with lamellar setae on the left and endopod on the right. □A. CN 115382, from Maotianshan, level M3, upper part, ×6. □B. CN 115383, from Maotianshan, level M3, lower part, × 6. □C. CN 115316, from Maotianshan, level M2, lower part. In this specimen the endo- and exopods are folded over each other, ×6. □D. CN 115315, from Maotianshan, level M2, lower part, ×6. □E. CN 115384, from Maotianshan, level M2, upper part, specimen showing medial sternites with annulated base of soft cormus leading to biramous appendage. The exopod is turned over forwards from its life posture, ×6.

Fig. 42 (right). *Naraoia longicaudata* Zhang & Hou, 1985. CN 115317, from Maotianshan, level M2, lamellar exopod setae from two successive appendages. Every individual seta tilts to the left, ×8.7.

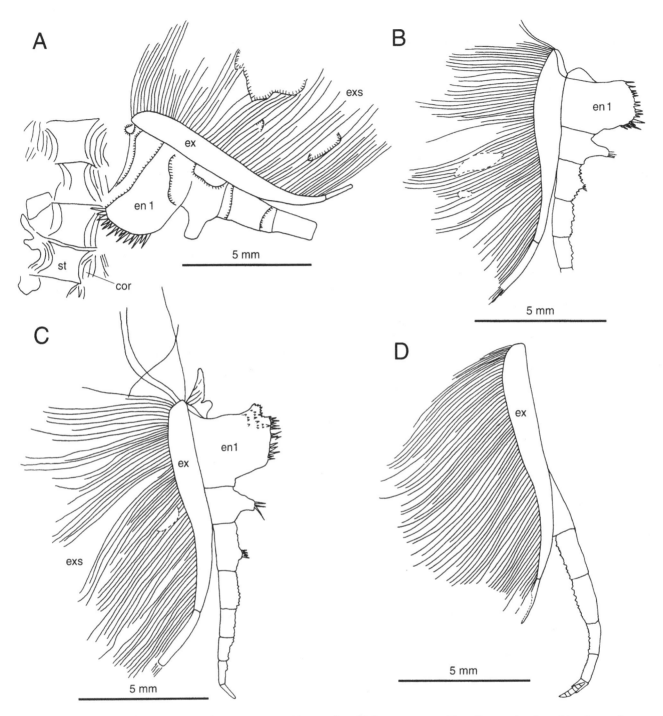

Fig. 43. Appendages of *Naraoia longicaudata* Zhang & Hou, 1985. □A. Drawing of CN 115384, same as Fig. 41E, appendage in position close to the row of sternites. Curved lines represent the ringed cormus proximal to the appendage proper. □B, C. Drawings of CN 115383 (B) and CN 115382 (C), representing flattened-out isolated limbs. Same as Figs. 41B and A. □D. Drawing of lower part of CN 115316, a specimen in which the two branches of the leg have become folded against each other. Same as Fig. 41C. For abbreviations, see Fig. 9.

Both *Naraoia* species are among the most abundant species in the Chengjiang fauna in terms of numbers of individuals. Disarticulated shields predominate, but there are hundreds of complete specimens of both species with soft parts that modify our view of *Naraoia* that is based on the Burgess Shale species *N. compacta* (see Whittington 1977).

In both Chinese species the gut is often seen to be filled with mud (Figs. 39D, 47). On the underside, a large, oval hypostome is located in the middle of the head. Anteriorly of the hypostome is a pair of small round projections, which may represent compound eyes (Figs. 39A–C, 40A, B). A single pair of antennae stem from each side of the hypostome, behind which, in both species, are three pairs

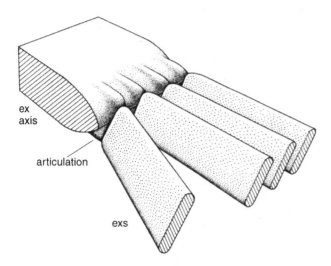

Fig. 44. Naraoia longicaudata Zhang & Hou, 1985. Attachment of lamellar setae on the exopod shaft. The setae are not vertical, but inclined towards the midline of the animal (to the left). For abbreviations, see Fig. 9.

of biramous legs in the head, as described from the Burgess Shale species. The body of *N. longicaudata* has 22 pairs of legs, that of *N. spinosa* about 15 pairs.

The limb structure, even in *N. longicaudata* (the most similar species to *N. compacta*), differs markedly from Whittington's (1977, Figs. 1, 96–99) reconstructions of *Naraoia compacta*. The proximal element is a basis. Whittington rotated this element, which he regarded as a coxa, some 45° to make it close against the underside of the

body. Our material shows that there was a wrinkled unsclerotized shaft (cormus) extending between the body and the basis (Figs. 41E, 43A). This flexible shaft made the leg very movable; in some specimens legs extend laterally, in others straight forwards or backwards. In Whittington's reconstruction, the exopod has a very short articulation with the basis (supposed coxa). Based on one of our specimens, Ramsköld & Edgecombe (1996, Fig. 1E) suggest that the hinge is longer, connecting the exopod with the entire length of the basis. Additional material, however, shows that the hinge is even longer, extending along the whole of the basis and (at least part of) endopod segment 1 (Figs. 41A–D, 43B–D). Also, each exopod 'filament' (Whittington's terminology) is articulated at its base. By definition, therefore, each 'filament' is a seta. The setae are flat (Fig. 42), but not because of sediment compaction, since they are inclined to the sedimentation surfaces in a consistent way: when a laterally extended leg is seen from above, the individual setae always dip fairly steeply towards the midline of the animal (Fig. 44). It is therefore easy to distinguish between the upper and lower surfaces of the leg. All setae are attached to the main element of the exopod shaft. A smaller distal element has fine bristles distally (Figs. 41A–B, 43B–C).

Whittington's reconstructions of *N. compacta* shows five endopod segments between the supposed coxa and the terminal element (Fig. 45A). Our identification of the supposed coxa as the proximal endopod segment changes the count to six (Fig. 45B). However, even this count is questonable, since *Naraoia longicaudata* has seven endopod segments proximal to a distal tarsus. In the latter,

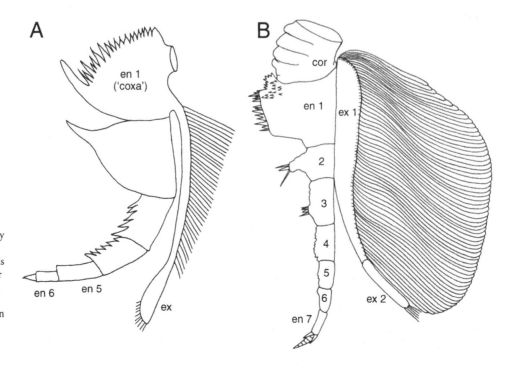

Fig. 45. Biramous appendage of *Naraoia.* □A. As reconstructed by Whittington (1977, Fig. 97) □B. As understood herein. The basis is more voluminous than the other podomeres, which appear to be flat. Therefore it suffered more distortion, and the reconstruction is only tentative. For abbreviations, see Fig. 9.

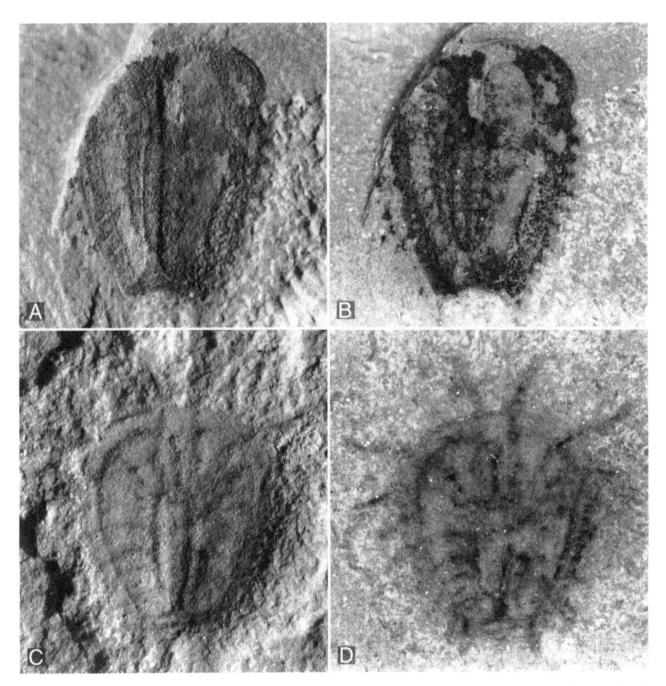

Fig. 46. Naraoia spinosa Zhang & Hou, 1985. Larvae. □A, B. CN 115284, from Jianbaobaoshan, level Dj1, lower part. Spine row visible along right side demonstrates that the larva belongs to this species, ×22. □C, D. CN 115281, from Maotianshan, level M2, lower part, exposing appendages. The relief photo (C) shows collapse of dorsum over appendages, while contrast photo shows remains of actual specimens, ×22. Panchromatic (A, C) and orthochromatic (B, D) film.

endopod segment 1 (the basis) has a large endite with spines on the lower surface. Endopod segment 2 has a smaller endite, drawn out in a point in the Burgess Shale species. The more distal segments lack endites in *N. longicaudata* (Figs. 41, 43).

When the limb is found detached, it is either spread flat or folded double along the long hinge between the two branches (Figs. 41, 43). This shows that the main move-

ment of the limb was in its proximal shaft, while the two branches were kept together.

Based on dorsal characteristics, the two Chengjiang species of *Naraoia* are quite similar to the two Burgess Shale *Naraoia* species. However, there is a more marked contrast between the appendages of the two Chinese species, since *N. longicaudata* has a narrow exopod shaft (such as described from *N. compacta*), while in *N. spinosa*

Fig. 47. Naraoia spinosa Zhang & Hou, 1985. □A, B. CN 115385, from Northwest slope of Maotianshan, level Cf1, which is the lowest bed of soft-bodied fossils and 20 m above the *Abadiella* bed. Lower part, complete individual showing marginal spines and intestinal diverticulae. On the right side of the thorax are seen three exopod blades, which are notably larger than the narrow exopod shaft in *N. longicaudata*. Panchromatic (A) and orthochromatic (B) film, ×3.7. □C. CN 115386, from Maotianshan, level M3, lower part, small specimen, ×12. □D. CN 115387, from Maotianshan, level M2, lower part, small specimen with wide mud-filled gut and diverticula, ×12.

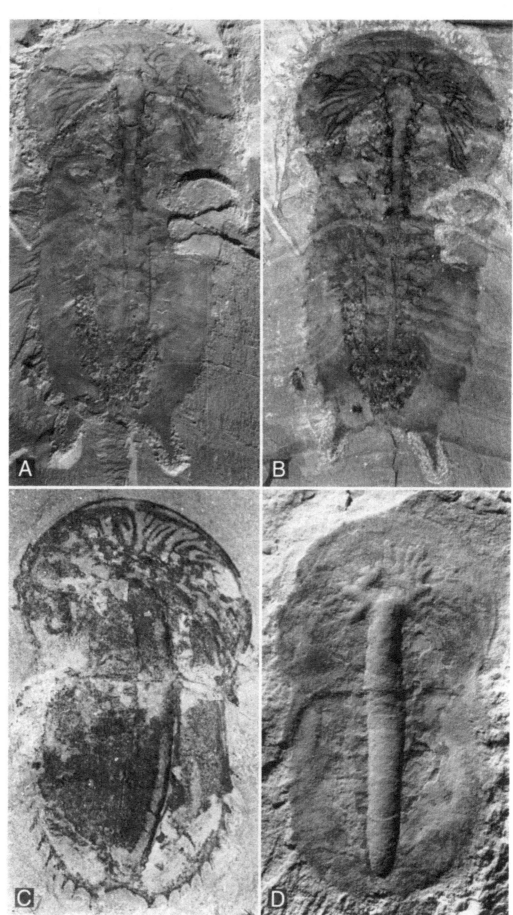

the exopod shaft is expanded into a broad flap (Fig. 47A–B). The attached setae appear to be shorter in *N. spinosa* than in *N. longicaudata*. The difference may reflect functional or ecological variability, but its systematic significance is not clear. We know almost nothing about the variation of appendages within genera in the Cambrian. At least for the moment, and also since we have not studied *N. spinosa* in detail, we leave *N. spinosa* within *Naraoia*.

At least in *N. longicaudata* there are clearly defined segmental sternites extending between the limbs (Figs. 41E, 43A). The lateral margin is excavated to fit the rounded shape of the leg shafts.

The smallest *Naraoia* specimens are considered to be protaspids, as they are similar to trilobite protaspids, and to belong to *N. spinosa*, as they have 11 pairs of marginal spines, including a large posterior pair, and have a posterior embayment in the posterior tergite (Fig. 46).

It has been suggested that the superficial similarity between the protaspis of *Naraoia* and the Vendian *Parvancorina* implies that the two are related (e.g., Conway Morris 1993). The supposed legs of *Parvancorina* extend not only from the longitudinal axis, but also from the ridge along the 'front' end of the body, from where they extend posteriorly. Furthermore, they appear to be part of the topography of the upper surface rather than something visible through a shield. In addition, many *Parvancorina* specimens definitely lack a perfect bilateral symmetry, as the anchor-shaped ridge is asymmetrical, and the margin outside the ridge is wider on one side than on the other. The evidence is sufficient to conclude that *Parvancorina* is no arthropod.

Order Retifaciida n. ord.

Diagnosis. – Lamellipedians with stalked ventral eyes, long slender tail behind tail shield, reticulate tergum in type genus, head with antennae and three pairs of appendages, and thorax of ten segments.

Family Retifaciidae n. fam.

Diagnosis. – As for the Order Retifaciida.

Genera included. – *Retifacies* Hou, Chen & Lu, 1989; possibly *Squamacula* n.gen.

Genus *Retifacies* Hou, Chen & Lu, 1989

Type species. – *Retifacies abnormalis* Hou, Chen & Lu, 1989

Diagnosis. – As for the family.

Retifacies abnormalis Hou, Chen & Lu, 1989

Figs. 48–52

Synonymy. – □1989 *Retifacies abnormalis* gen. et sp. nov. – Hou *et al.*, pp. 53–54, Pls. 1:1–6; 2:1–4; Text-figs. 1–2. □1990 *Tuzoia* sp. – Shu, p. 89, Pl. 2:4. □1991 *Retifacies abnormalis* Hou *et al.* – Hou & Bergström, p. 183. □1991 *Retifacies abnormalis* Hou, Chen & Lu – Delle Cave & Simonetta, p. 201, Fig. 6B.

Holotype. – CN 108298 (Hou *et al.* 1989, Pl. 3:1), a complete specimen lacking appendages, from Maotianshan, level M3.

Other specimens. – Six paratype specimens (CN 108299–108304) were illustrated by Hou *et al.* (1989). Specimens CN 108299, 108301a–b, and 108303a–b are completely preserved exoskeletons, whereas the other three (CN 108300, 108302 and 108304) consist of isolated tail shields. An isolated tail shield from the same locality and level as those from Chengjiang was erroneously illustrated by Shu (1990, p. 89, Pl. 2:4) as a valve of *Tuzoia*.

New illustrated specimens include the following: CN 115388 (Figs. 48A, 51) is a complete specimen with well-preserved appendages. CN 115389 (Fig. 48B) and CN 115390 (Fig. 49A) are complete exoskeletons. CN 115392 (Fig. 49B) is an isolated tail shield. CN 115391 (Fig. 48C) is an isolated fragmentary head shield.

Distribution. – Maotianshan, levels M2, M3, and Cf6, Jianbaobaoshan, level Dj1, and Xiaolantian, level XL1. Most specimens are from M2.

Description.. – (a) General characteristics: The general outline is oval. The dorsum is divided into head and tail shields and ten thoracic tergites. There is a distinct overlap between neighbouring tergites. The underlapping anterior border forms a raised band over the posterior main area. The tergum is pressed flat but appears to have been smoothly vaulted, particularly in its axial part. There is no evidence of axial furrows. The width of the axis is about ¼ of the width of the animal. The exoskeletal surface is ornamented with an irregular network of large, polygonal meshes. A narrow doublure extends all around the exoskeleton. The tail is long, slender, and segmented. Two complete specimens are 5.3 cm long and 3.2 cm wide (Fig. 48A) and 6.1 cm long and 3.7 cm wide (Fig. 49A) respectively. Half or less of a head shield (Fig. 48C) is 12 cm wide. This means that the head could be at least 20 cm wide, the entire tergum some 35 cm long, and the animal inclusive of antennae and tail at least 55 cm long.

(b) Head: The head shield is devoid of characteristic features, save for the reticulation. The meshes are irregularly pentagonal or hexagonal and smaller than on the thoracic and tail tergites. There are no dorsal eyes. The anterior margin is smoothly curved and the posterior

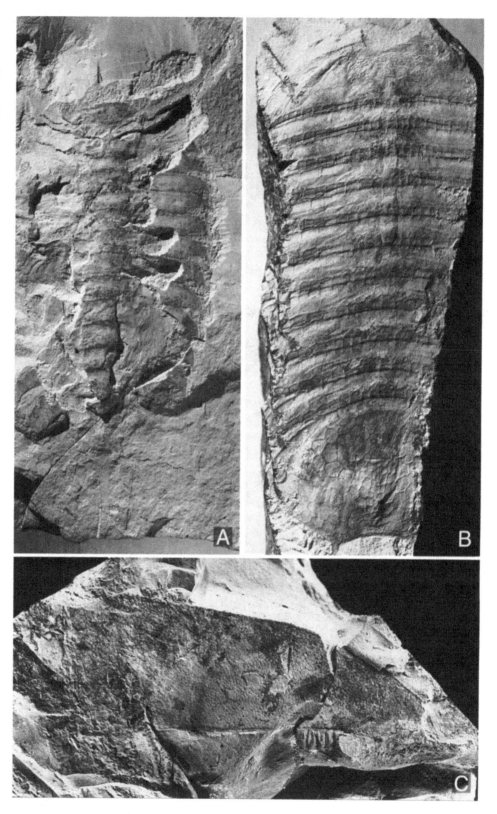

Fig. 48. Retifacies abnormalis Hou et al., 1989. □A. CN 115388 from Maotianshan, level M2, lower part, prepared to show appendages. In addition, long tail spine is exposed, ×1.7. Compare drawing Fig. 51. □B. CN 115389, from Maotianshan, level M2, lower part with nice ornament, ×3. □C. CN 115391, from Maotianshan, level M3, upper part, fragmentary head shield, ×0.85.

margin straight. The head width is about 4–4.5 times greater than the length. Impressions of appendages are occasionally seen (Fig. 48A).

(c) Body: The thorax is of virtually constant width. From the 5th or 6th tergite, the widest portion of the body, there is a slight narrowing both forwards and backwards (Fig. 49A). The first five thoracic tergites appear to be progressively longer, whereas the succeeding ones are of equal length (Figs. 48B, 49A). The pleural spines are short, pointed and of similar morphology (Figs. 48A, 51).

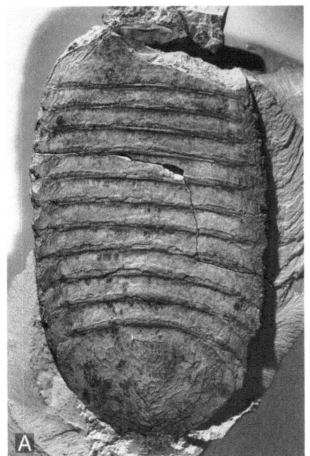

Fig. 49. Retifacies abnormalis Hou *et al.,* 1989. □A. CN 115390, from Maotianshan, level M2, lower part, photographed with red filter. Dorsal side with ornament, and exposed hypostome, ×1.7. □B. CN 115392, from Xiaolantian, level XL1, lower part, isolated tail. This is tilted forwards and the margin therefore preserves much of the dorsal convexity, ×2.5.

In the pleural region, each thoracic tergite has an anterior and a posterior row of ornamental meshes. The anterior meshes are large and irregularly rectangular, the posterior ones small and approximately square. The anterior row covers the middle band of the tergite, behind the

anterior raised band, whereas the posterior row covers the part of the tergite that overlaps the successive tergite. In the axial region the meshes are smaller (Fig. 48B).

The tail shield is wide and large, elliptical in outline, bearing a pair of short, posterior spines. The underlapping anterior band is clearly shown on the isolated tail shield; it is smooth and bounded posteriorly by a narrow ridge. The amount of overlap appears widest laterally (Figs. 48B, 49B; Hou *et al.* 1979, Pls. 1:3, 6; 2:4).

(d) Ventral side: The narrow doublure widens anteriorly into a hypostome with smoothly curved outline (Figs. 48A, 49A, 51). The hypostome appears to have an even marginal outline. The hypostome extends back to the supposed mouth region below the posterior part of the head shield.

A pair of antennae is well preserved in one specimen (Figs. 48A, 51). The right antenna is complete in the lower part; the left one is preserved in its upper part, showing annuli and bunches of 5–7 setae. The setae are inserted near the annular boundaries along the inner side; they have not been seen along the outer, dorsal or ventral sides. In the two specimens with antennae, the antennae seem to originate from the sides of the hypostome (Figs. 48A, 51). The base of the left antenna seems to extend inward to the lateral embayment of the hypostome, suggesting that antennae originate from there (Figs. 48A, 51). The antenna has about 25 annuli. The proximal 3–4 annuli are approximately twice as wide as they are long, making the antenna look sturdier than in most other early arthropods. The succeeding annuli are longer and slimmer, the middle and distal ones being approximately twice as long as wide (Fig. 51). A pair of short, club-like structures occur near the base of (probably in front of) the antennae (Figs. 48A, 51); each club is thickest distally and progressively thinner toward its base. The club-shaped organ is probably a stalked eye.

There are three pairs of biramous cephalic limbs behind the uniramous antennae. They are notably more closely spaced than the successive limbs, of which there are ten thoracic pairs and five pairs in the tail. Ringed impressions reveal the presence of a basal soft cormus between the body and the proximal podomere of each leg. The right leg of the head and the 5th left and 7th right legs on the thorax were exposed by preparation. The outer branch of the second left tail leg was well exposed by a natural split. All biramous limbs have the same structure and show only size differentiation (Fig. 48A, 51). The legs of the head and anterior part of the thorax tend to be rotated forward at an angle of 90°, so that the endopod is placed in front of the exopod and the ventral side of endopod is directed anteriorly. The right 7th thoracic leg is slightly removed from its site of attachment, shifted a little outward and rotated backwards and inwards. This leg clearly exposes the structure of the exopod, the articulating relationship between exopod and endopod, and the proximal

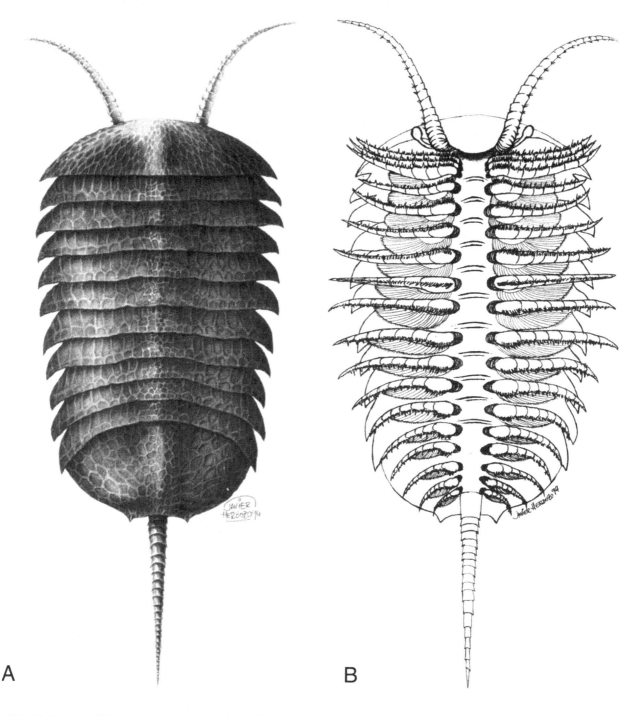

A　　　　　　　　　　　　　　　　　　　　　　B

Fig. 50. Retifacies abnormalis Hou *et al.*, 1989. Reconstruction in dorsal and ventral views. Note reticulation of dorsum, with difference in pattern between head, thorax and tail, and between axial and pleural regions. Note also ventrally positioned paired eyes and crowding of appendages in head.

structure of the endopod. The shaft of the exopod is composed of an oval flap. The articulating margin of the flap is as long as both basis and endopod podomere 1 (Figs. 51–52) and is completely hinged with them. This articulation relationship is also suggested by the 3rd right cephalic limb. The posterior margin of the exopod flap has a regular row of about 20 flat and comparatively broad setae, each articulated at its base. The setae form an imbricated series, with each member tilted toward the midline of the animal. On the 5th left thoracic limb the dorsal edge of each seta carries some seven or eight spine-like bristles.

Between the basis and the body is a short series of curved wrinkles indicating (as in *Naraoia longicaudata*) a proximal, soft-skinned cormus. Between the limb pairs appears to be a series of segmentally arranged median

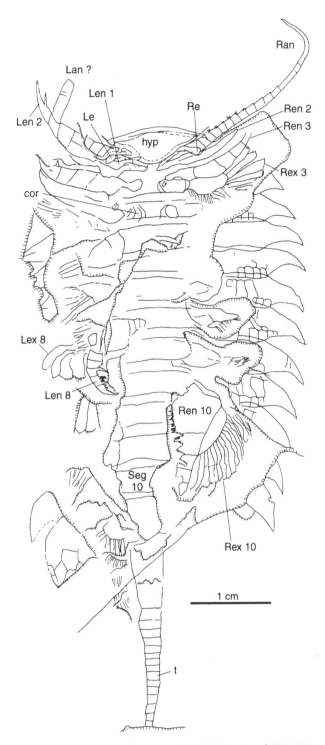

Fig. 51. Retifacies abnormalis Hou *et al.,* 1989. Drawing of CN115388 after preparation. On both sides of the hypostome are a pair of antennae and a pair of small, club-shaped organs, which apparently represent stalked compound eyes. Curved lines at the base of a few appendages represent the annulated soft cormus proximal to the skeletonized part of the appendage. For abbreviations, see Fig. 9.

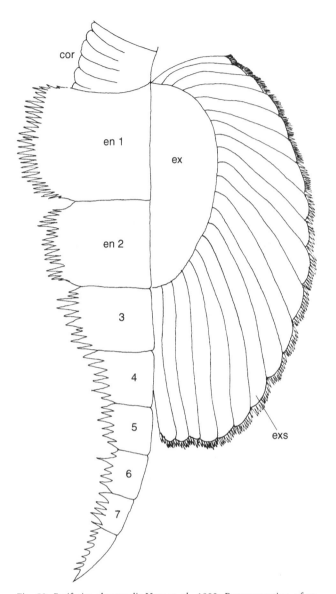

Fig. 52. Retifacies abnormalis Hou *et al.,* 1989. Reconstruction of an appendage, showing the annulated soft cormus, endopod and exopod. There is an articulated contact between the exopod and 1st two podomeres (basis and 1st post-basis segment). The tilted exopod setae are even broader than they appear from the drawing since they overlap each other. For abbreviations, see Fig. 9.

sternites. The proximal part of the endopod is well shown in the 7th right thoracic limb, whereas the distal part is best seen in the 5th left thoracic limb. The basis is succeeded by probably seven podomeres including the termi-

nal piece, which is notably long (indicating it to be a true podomere rather than just a terminal spine). The basis is large and has at least 30 spines on the ventral side. Each of the succeeding podomeres also carries not less than five ventral spines (exact number not known). The leg becomes successively narrower distally (Fig. 52).

A long, spine-like, but segmented or annulated tail extends backward from under the tail shield (Figs. 48A, 50, 51). A partial sediment infilling of the intestine can be seen in the anterior half of one specimen (Fig. 48B).

(e) Discussion: Delle Cave & Simonetta (1991, p. 201) regard the visible plate as a probable hypostome; this in their view indicates an affinity with trilobites. They regard

Retifacies as belonging to the *Kuamaia–Helmetia* group. However, the apparent absence of a rostral plate, the ventral position of the eyes, and the overlap between successive tergites are arguments against such a conclusion.

Delle Cave & Simonetta (1991, p. 201) mention a maximum length of about 75 mm for *R. abnormalis*, and therefore regards the species as a comparatively large animal. As mentioned above, our material indicates a maximum length of at least 350 mm, antennae and tail not included.

Family ?Retifaciidae

Genus *Squamacula* n. gen.

Name. – Latin *squama*, scale; *-culus*, diminutive suffix; referring to the outline of whole animal in dorsal view.

Type species. – *Squamacula clypeata* n. gen. et sp.

Diagnosis. – An unornamented genus of ?Retifaciidae with very broad body.

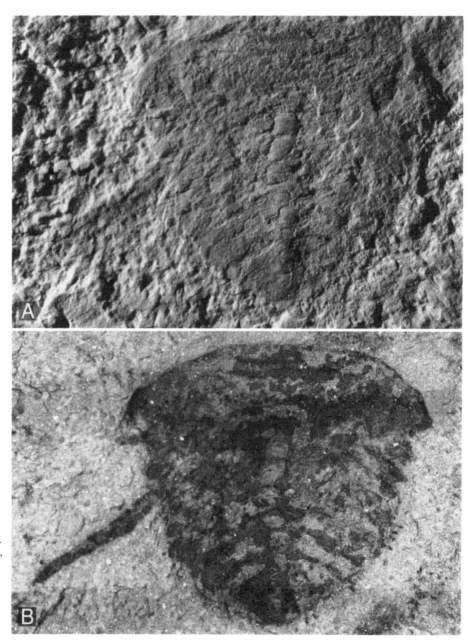

Fig. 53. Squamacula clypeata n. gen. et sp. CN 115394, from Xiaolantian, level XL1, lower part. A long antenna extends on the left side. Mud-filled gut particularly well seen in the upper figure. Panchromatic (A) and orthrochromatic (B) film, ×12.

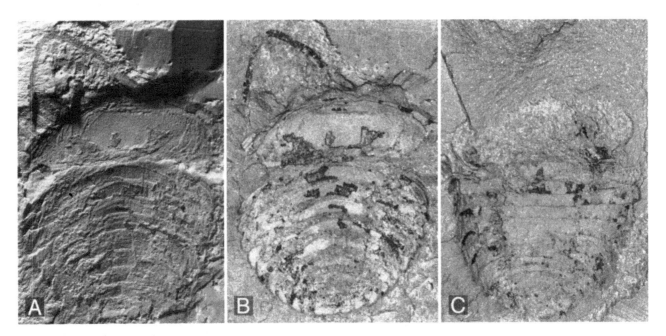

Fig. 54. Squamacula clypeata n.gen. et sp. □A, B. Holotype, CN 115393, from Maotianshan, level M2, upper part, showing a long antenna beyond the head shield. This specimen demonstrates the original convexity through the V-shaped form of the segmental tergites that are tilted forwards, and also through the wrinkling of the head shield and tergites, panchromatic and orthochromatic films respectively, ×8. □C. CN 115395, from Maotianshan, level M2, upper part, showing fragmentary endopod and exopod in the upper right, ×8.

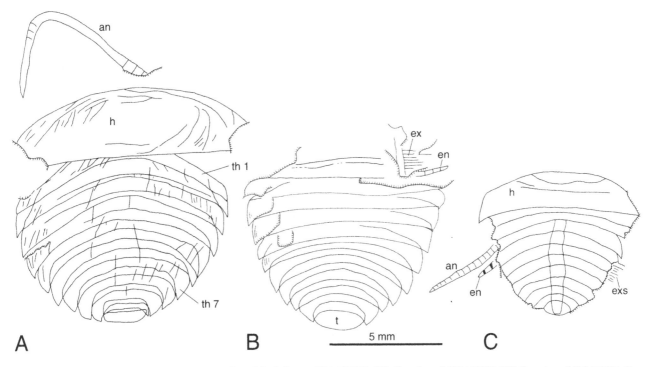

Fig. 55. Squamacula clypeata n.gen. et sp. □A. Drawing of the holotype (CN 115393). □B. Drawing of CN 115395. □C. Drawing of CN 115394. For abbreviations, see Fig. 9.

Discussion. – Although reminiscent of *Retifacies* in having broad head and tail shields and ten thoracic tergites, this genus is quite distinctive in its unornamented surface, rounded outline in dorsal view, and longer antennae with more annuli.

Squamacula clypeata n.gen. et sp.

Figs. 53–56

Name. – Latin *clypeatus*, shield-shaped; referring to the outline of the animal in dorsal view.

Fig. 56. Reconstruction of *Squamacula clypeata* n.gen. et sp. in dorsal view.

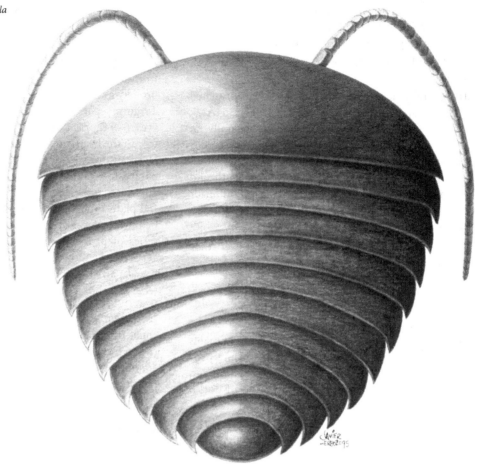

Holotype. – CN 115393, from Maotianshan, level M2, a specimen with fragmentary lower part and complete upper part (Fig. 54A–B). Legs are preserved in the lower part, but details are unclear.

Other specimens. – Two complete specimens with appendages more or less well preserved. CN 115394 (Fig. 53, a lower part) and CN 115395 (Fig. 54C, an upper part), respectively, both with lower and upper parts.

Diagnosis. – As for the genus.

Distribution. – Maotianshan, level M2, and Xiaolantian, level XL1.

Description. – (a) General characteristics: The animal is small, rounded in outline, broader than long. All three specimens are dorsoventrally flattened; the largest is 10.2 mm long and 10.2 mm wide (Fig. 54A–B), the smallest 6.3 mm long and 7.1 mm wide (Fig. 53). The exoskeleton, which is composed of a head shield, 10 thoracic tergites and an elliptical tail shield, is strongly vaulted but lacks an axis defined by furrows. The surface appears smooth but has irregularly placed wrinkles owing to compression of the convex exoskeleton (Figs. 54A–B, 55A). The overlap between adjoining thoracic tergites amounts to about a third of the tergite length (Figs. 54, 55A–B). To some extent this great overlap may be caused by compression. A convex filling of the intestine is shown in one specimen (Fig. 53, 55C), extending from the third thoracic tergite to near the end of the tail shield.

(b) Head: The head shield is wide and short, width:length ratio about 4. The anterior margin is smoothly curved (Figs. 53, 54A–B, 55A, C). The posterior margin is probably straight (Figs. 54A–B, 55A); in one of the specimens (Figs. 53, 55C) it is slightly curved forward, probably owing to a slight forward tilt of the head prior to compression. The genal spines are very short, pointed and directed downwards (Figs. 53, 55C).

(c) Body: The body narrows backwards progressively from the first thoracic tergite, forming half an ellipse. Each pleura ends in a very short spine. The anterior (outer) side is so curved that the spines point backwards in the anterior part of the body but backwards–inwards more posteriorly (Figs. 54, 55A–B). There are ten thoracic tergites, all of similar length. In one specimen, the anterior tergites appear straight, while the more posterior ones are distinctly curved (Figs. 54C, 55B); in other specimens, the anterior tergites arch forwards in the middle. This difference is obviously due to a difference in tilting, as demonstrated in the holotype by the fact that head and thorax are separated and show quite different curvatures at the break (Figs. 54A–B, 55A). The small tail shield is elliptical and is embraced on three sides by the last thoracic tergite.

(d) Ventral side: In the holotype, a long annulated structure occurs in front of the head shield (Figs. 54A–B, 55A). A similar structure extends posterolaterally in another specimen (Figs. 53, 55C). These presumed antennae equal the entire tergum in length and are fairly stout, having an estimated 35 annuli with similar length and width. In one specimen, a transverse line over the anterior portion of the head shield may indicate the margin of a hypostome (Figs. 53, 55C).

All specimens are preserved with legs. However, they are strongly weathered, and the details are destroyed. Individual segmented endopods can be seen, as well as exopod setae (Figs. 55B–C).

Discussion. – A reconstruction (Fig. 56) indicates the characteristics of the animal in dorsal view. This animal is insufficiently preserved to serve as a basis for any prolonged discussion. It resembles *Retifacies abnormalis* in having a short, very broad head shield devoid of dorsal eyes, ten thoracic tergites of similar shape, an oval tail tergite, and sturdy antennae. It differs considerably from *R. abnormalis* in shape, being more convex and much shorter, and in lacking ornament.

Subclass Conciliterga n.subcl.

Name. – Latin *concilio*, to unite separate parts into a whole, and Lat. *tergum*, back; referring to the fused tergal shield.

Diagnosis. – Artiopodans with semitergites, i.e. incompletely separated tergites that overlap only in the axial region, if at all, and may be more or less fused to form a single shield covering the entire body. When dorsal compound eyes are present, the visual surface is directed upwards. Consistent lack of gut filling (which indicates that feeding did not involve mud ingestion).

Order included. – Order Helmetiida Novozhilov, 1969.

Discussion. – The semi-tergites represent a rare construction. The edge-to-edge connection at the margin of the shield indicates that overlapping movements were not possible there, although some bending may have occurred. The stepped appearance in the middle of the shield indicates the possibility of a very limited flexure by means of the contraction of longitudinal muscles. On the whole, however, the shield must have been inflexible, particularly since it was vaulted.

Order Helmetiida Novozhilov, 1969

(=Order Helmetida Simonetta & Delle Cave, 1975)

Emended diagnosis. – Exoskeleton with flat sides sloping from a hogback-like midline without defined axial region, broadly to narrowly oval in outline, with smooth surface. Tergum separated into head shield with pair of sessile eyes, 6–9 thoracic tergites, and a large tail shield. Anteriorly in the head, one transverse and a pair of longitudinal facial sutures define a small rostral plate and a pair of pararostral plates. On the ventral side, a hypostome appears to be situated just behind the doublure of the rostral plate. One pair of dorsal compound eyes placed just behind the level of the rostral plate. From head to tail, a series of probably uniform biramous appendages with very coarse endopods. Antennae and probably three additional pairs of limbs in head, one pair beneath each of tergites, probably at least four pairs beneath tail shield.

Discussion. – Characteristic features include a head with a small rostral plate inserted anteriorly in the head shield (i.e. in front of an anteriorly placed transverse suture) and adjoining pararostral plates, a moderate number of thoracic tergites (6–9, as far as known), and a large tail shield extended into a posteromedian spine and one to several pairs of lateral spines. Tergites do not overlap laterally but meet edge to edge, and they may also be fused into a single shield. There is some indication of mobile junctions in the middle where there is a short overlap, but on the whole the tergum appears to have been fairly rigid.

The only previously known form was *Helmetia expansa* Walcott, 1918, from the Burgess Shale. There is a lot of confusion in the literature on this species. As far as we know, *Helmetia* differs from the forms described herein mainly by the possession of anterolateral horns, which are connected with a lateral elongation of the pararostral plates seen in the Chinese members of the group. The appendages of *H. expansa* are poorly known and were thought to be represented only by exopodal setae (Walcott 1931, Pl. 23). Nevertheless, their disposition made it possible for Størmer (1944 and *in* Moore 1959) to conclude that *Helmetia* is a member of the Trilobitomorpha. Størmer (1944, p. 135) believed that the endopods ('telopodites') were reduced. Actually, the holotype there appears to show fragmentary remains of the endopods on the left side as well as unmistakable evidence of three (possibly four) successive exopods in the head (Simonetta & Delle Cave 1975, Pl. 16:1); Simonetta & Delle Cave state that there were probably two pairs, whereas their reconstruction shows one pair of legs in the head (Simonetta & Delle Cave 1975, p. 3 and Pl. 2, respectively, and Delle Cave & Simonetta 1991, p. 201 and Fig. 6F, respectively). The tail shield covers at least five successive exopods, but only two pairs are included in Delle Cave & Simonetta's reconstruction.

Helmetia expansa has been reconstructed without eyes (Størmer 1944, Fig. 17:8–9; Størmer *in* Moore 1959, Fig. 25), or with a pair of ventrally placed eyes (Simonetta & Delle Cave 1975, Pl. 2:2 and p. 3; Delle Cave & Simonetta 1991, Fig. 6F). Actually, the holotype clearly shows a pair of round structures on the dorsal side of the head shield, just where the eyes are expected from knowledge of the Chinese helmetiids.

The rostral plate has been mistaken for a misplaced hypostome (Størmer 1944, p. 88; Simonetta & Delle Cave 1975, p. 3).

Størmer (1944, p. 88) thought that *Helmetia* was related to *Mollisonia* Walcott, 1912, and *Tontoia* Walcott, 1912. Whittington (1985b) mentioned *Helmetia* only in his list of species, where it is classified only as an arthropod. In the position of an anterior head suture passing the eyes, the Helmetiida are similar both to the Xandarellida and to the Limulavida, the latter preserving an open joint in the same position.

Families. – Helmetiidae Simonetta & Delle Cave, 1975, Tegopeltidae Simonetta & Delle Cave, 1975, Skioldiidae n.fam., and Saperiidae n.fam.

Family Helmetiidae Simonetta & Delle Cave, 1975

Diagnosis. – As for the order.

Genera included. – *Helmetia* Walcott, 1918, *Kuamaia* Hou, 1987(b), and *Rhombicalvaria* Hou, 1987(b).

Of the two genera that are distinguished on Chinese material, *Rhombicalvaria* is represented by a rare species, and only *Kuamaia* is treated in detail below.

Genus *Kuamaia* Hou, 1987

Type species. – *Kuamaia lata* Hou, 1987

Diagnosis. – Head shield sub-trapezoidal in outline, with straight anterior margin, rounded anterolateral angles, and spiny posterolateral angles; 7–8 thoracic tergites with short pleural spines; tail shield with two or three pairs of marginal spines similar to the pleural spines.

Discussion. – *Kuamaia* closely resembles *Helmetia* Walcott, 1918, from the Burgess Shale in its trapezoidal head shield and spiny tail shield. It differs in the straight rather than backwardly curved anterior margin of the head shield, rounded rather than pronouncedly spiny antero-lateral angles, and spiny rather than pointed pleural extremities. The significance of the difference in numbers of thoracic segments is unclear. The differences between *Kuamaia* and *Rhombicalvaria* are much more subtle, and

these two taxa could easily be considered as synonyms (Delle Cave & Simonetta 1991, p. 201). As described, *Kuamaia* has 7–8 thoracic segments and 2–3 pairs of tail spines, *Rhombicalvaria* 9 and 1, respectively, and much more extended spines. It is questionable if these differences merit generic distinction, but in the present context we treat them as separate.

Species. – *Kuamaia lata* Hou, 1987, and *Kuamaia muricata* n.sp.

Kuamaia lata Hou, 1987

Figs. 57–60

Synonymy. – □1987b *Kuamaia lata* gen. et sp.nov. – Hou, pp. 283–284, Pls. 1:4; 3:1–2; 4:1–7; Text-fig. 3–4. □1991 *Kuamaia lata* Hou – Hou & Bergström, p. 183. □1991 *Kuamaia lata* Hou – Delle Cave & Simonetta, p. 201, Fig. 6E.

Holotype. – CN 100128 (Hou 1987b, Pl. 23:1–2) from Maotianshan, level M2, an incomplete specimen including upper and lower parts.

Other specimens. – Six paratypes (CN 100129–100134) were illustrated by Hou (1987b), all incomplete dorsal exoskeletons. This description is furthermore based on CN 115318, 115396–115400 (Fig. 57) and about 30 additional specimens.

Distribution. – Most of the specimens are from Maotianshan, level M2. A few came from the northwest slope of Maotianshan, level Cf5, from Jianbaobashan, levels Dj1 and Dj2, and from Xiaolantian, level XL1.

Diagnosis. – A *Kuamaia* species with seven thoracic tergites and two pairs of lateral spines on the tail shield.

Description. – (a) General characteristics: Of the about 40 specimens, most are fragmentary and only two have partly preserved legs. The exoskeleton is dorsoventrally compressed. The outline ranges from elongately to broadly oval and is always more or less symmetrical, indicating burial parallel to the bedding plane.

Although the specimens are much flattened, the midline is in some specimens raised into a low rounded ridge (Figs. 57A–B, G, 58A), indicating that the dorsal profile was originally like that of a low roof with flat, sloping sides (Fig. 60). The unmineralized exoskeleton is preserved as a thin white film or shows a purplish red colour.

A thin raised band along the lateral margins of the head shield, pleurae, and tail shield suggests that there was a narrow doublure (Fig. 57A, C, F).

Sagittal length of five complete specimens: 31.5–59 mm; maximum width: 19.5–38.7 mm.

(b) Head: The head shield is broad, short. The width:length ratio is about 3; the sides are rounded. There

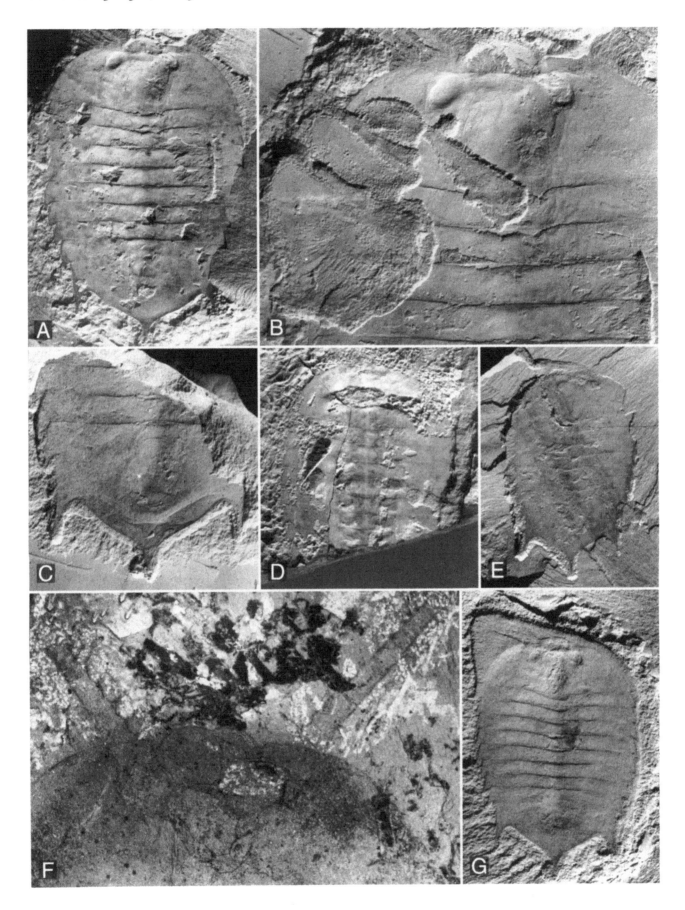

are very short genal spines. Anteriorly, a transverse facial suture just in front of the compound eyes marks the boundary between the main part of the head shield, a rounded median rostral plate, and an adjoining pair of narrow pararostral plates (Figs. 51A, B, 58A, 59A). The rostral plate extends forwards tongue-like. Paired sutures

also separate the rostral plate from the pararostral plates. It is not known if the so-called 'sutures' functioned during moulting.

The posterior margin of the head shield is preserved as a more or less straight or curved line, depending on the tilt of the head shield during embedding in the sediment.

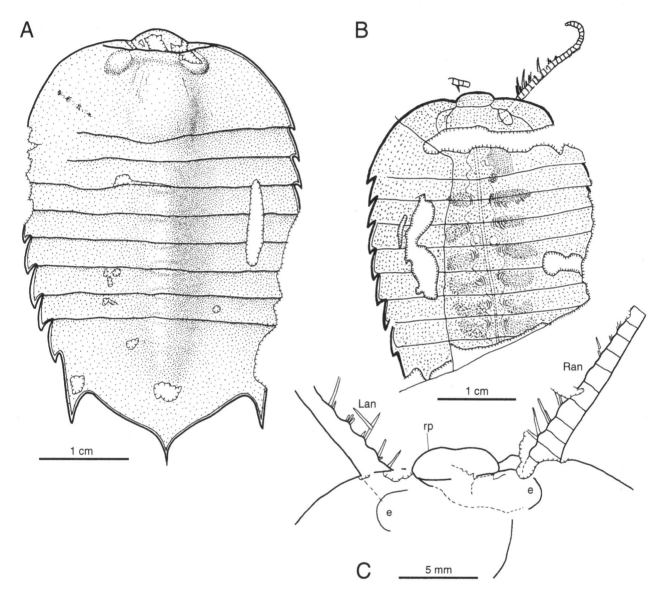

Fig. 58. Kuamaia lata Hou, 1987. □A. Drawing of CN 115318 before preparation. Note a segmental depression on the left side of the head shield. Preparation revealed an endopod in the depression; cf. Fig. 59A. □B. Drawing of CN 115397. Note paired antennae and their attachments, and curved lines similar to those of *Naraoia* and *Retifacies* and indicating the soft cormus of the appendage. □C. Drawing of anterior part of CN 115399, showing bristles on antennae. For abbreviations, see Fig. 9.

Fig. 57 (opposite page). *Kuamaia lata* Hou, 1987. In dorsal view. □A, B. CN 115318, from Maotianshan, level M2, lower part, before preparation and after exposure of left side appendage. Note also rostral and pararostral plates in front of eyes. A, with green filter, ×1.5, and B, coated with ammonium chloride, ×3.3. □C. CN 115396, from Maotianshan, level M2, lower part, showing convex axial area in the tail shield with one segmental boundary indicted by an impression, ×2.4. □D. CN 115397, from Maotianshan, level 2, lower part. Note impression of medial sternites with lateral wrinkled ends, much as in *Naraoia*. Green filter, ×1.6. □E. CN 115398, from Maotianshan, level M2, lower part, ×1.5. □F. CN 115399, from Maotianshan, Level M2, lower part, anterior part of a complete specimen, ×4.5. □G. CN 115400, from Maotianshan, level M2, lower part with a narrow axial area, ×1.5.

One pair of large, ovate compound eyes are present anteriorly on the head shield, where they rise above the surrounding area (Figs. 57A–B, G, 58A–B, 59A). There is no circum-ocular suture. In some specimens, the eyes are flattened through compaction (Figs. 57F, 58C).

In one specimen (Fig. 57A), the head rises more notably in the midline than in the other specimens, giving a faint impression of an axial area. On both sides of the midline there are thin and short wrinkles, extending slightly obliquely to the exsagittal line and continuing to tergites 1 and 2. Fractures are present on the convex rostral plate and right eye. In another specimen (Fig. 57G), the convex axial region is adequately defined by adjoining faintly concave areas. The external surface of the head shield is smooth.

(c) Body: The body is covered by seven thoracic tergites and a tail shield. The thoracic tergites are of virtually equal length. Tergite no. 4 is widest, the body smoothly tapering forwards and backwards (Figs. 57A, G, 58A, 59A). In different specimens, the first several tergites are preserved arching (tilting) backwards to a variable degree, while the posterior two or three tergites appear straight (Figs. 57A, G). The lateral spines increase in size from the head shield to the third tergite, beyond which their length is constant.

Neighbouring tergites overlap only axially, while they meet edge-to-edge laterally (Fig. 57A, G). The vertical step beetween the surfaces of adjoining tergites is largest in the midline and decreases laterally, so that it is present in only about ¼ of the animal width. The medial overlap shortens in a posterior direction, so that the overlap between tergite 7 and the tail shield is the shortest one.

The tail shield is broad and large, even longer (excluding the end spine) than the head shield, triangular in outline, bearing one terminal spine and two pairs of lateral spines. The anterior spines are much shorter than the posterior pair and are of the same length as the pleural spines of the posterior thoracic tergites. The posterior paired spines are approximately equal in length to the terminal spine (Figs. 57A, C, E, G, 58A, 59A).

(d) Ventral side: The supposed doublure of the rostral plate forms the anterior end. Paired antennae project along the sides of the rostral doublure beyond the anterior margin of the head shield (Fig. 57D, F, 58B–C). The antennal annuli are shorter than wide proximally, then become about equal in length and width, bearing a bulge with a bunch of 3–4 long setae on the inner side of each annulus. Specimen CN 115397 (Figs. 57D, 58B) shows an almost complete right antenna divided into annuli and carrying long setae. Beyond the margin of the head, some 27–30 annuli are seen; the total number is supposed to be around 35. Three or four setae are clustered on the inner side of each annulus (Figs. 57F, 58B–C). The proximal part of the antenna is represented by a clear impression on the right side of the head shield. A symmetrically placed furrow shows the position of the left antenna. A fragment

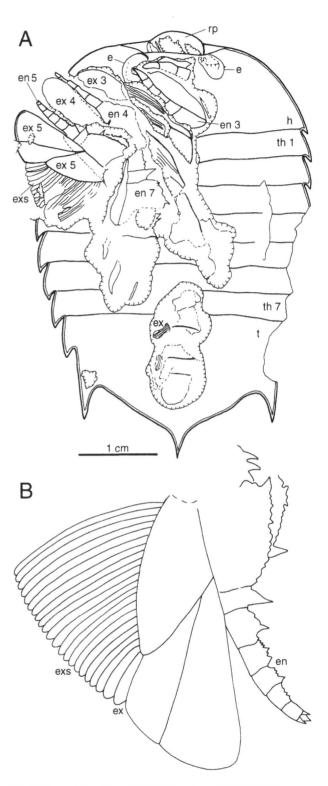

Fig. 59. Kuamaia lata Hou, 1987. □A. Drawing of CN 115318 after preparation. The right legs extend over the midline. □B. Reconstruction of biramous appendage. The most proximal portion poorly known. For abbreviations, see Fig. 9.

outside the rostral plate may represent a portion of the left antenna (Figs. 57D, 58B). There is indication that the antenna is attached to the side of a large oval structure situated just behind the rostral plate doublure (Figs. 57D,

Fig. 60. Reconstruction of *Kuamaia lata* Hou, 1985, in anterolateral view.

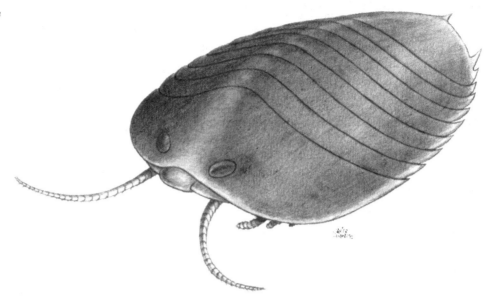

58B) and delimited by a curved shallow furrow, outside which the eye pair is shown. The position of the structure indicates that it may be a hypostome. The possible hypostome is seen also in specimen CN 115399 (Figs. 57F, 58C).

In specimen CN 115318, one endopod is indicated on the dorsum as a partly segmented furrow (Figs. 57A, 58A). Additional limbs are exposed under the head shield and thoracic tergites (Figs. 57B, 59A). The exposed limbs beneath the head shield appear to be biramous right limbs that are extended over the midline. They may be either bent over the midline or removed from their original place. The anterior pair of biramous legs is close to the rostral plate, while the posterior pair is close to the posterior margin of the head shield. There appears to be a fragment of a limb between them, seemingly representing the 2nd right limb under the head. Three pairs of biramous legs may thus be present beneath the head shield. Preparation failed to expose the hypostome.

In specimen CN 115318, preparation exposed exopod setae whose impressions were first seen on the dorsal side of the thoracic tergites (Figs. 57A–B, 59A). Each tergite corresponds to one pair of biramous legs. Three left limbs under the anterior tergites are well preserved; they may represent 1st (en4 and ex4 in Fig. 59), 2nd (en5, ex5) and 4th (en7, ex7 in Fig. 59) left thoracic limbs. The 1st and 2nd left limbs show a normal posture in extending forward and outward, while the 4th limb extends so strongly forwards that it parallels the sagittal axis. The biramous legs are spread flat, unfolded on the hinge between endopod and exopod. This hinge extends along the whole of both basis and the next endopod podomere, possibly even along part of a 3rd podomere (Fig. 59A–B). The endopod is composed of basis, six additional podomeres and a strong terminal spine flanked by two smaller spines (Fig. 59B). The basis is notably large and bears a strong endite

with two parallel rows of spines. The podomere distal to the basis seems to be short and has a stout enditic spine. The next four podomeres appear to be equal in length; the most proximal one has a short endite with a short spine, the others are devoid of endites but carry many small spines on the inner side. The main element of the exopod (Fig. 59B) is widest in its middle part and has a regular row of flat setae, each of which has a basal articulation. The distal exopod element is very large, wide, with a curved distal end. It has a long ridge that extends from the hinge line to the distal curved margin.

Another specimen (CN 115397) lacks most of the tail shield (Figs. 57D, 58B) and has a broad transverse gap in the head shield. An irregular, convex, axial region extends backward from the gap. Irregular ridges extend, pairwise, a short distance laterally from the midline. Two pairs of ridges are present on the head shield and one pair on each thoracic tergite. One pair is probably missing because of the damage in the head shield, making a likely total of (at least) three pairs in the head. As the structures are most probably connected with the limbs, this may mean three pairs of legs in the head. On the right side of the head, adjoining the lateral ridges, are imprints of exopodal setae, which may represent the right 2nd head exopod.

Between each two successive lateral ridges in specimen CN 115397 is a slightly depressed surface extending less than half the width of the pleura (Figs. 57D, 58B). The proximal part is commonly wrinkled, presumably because of compaction of the proximal parts of the legs. The depressed surface seems to retain impressions of exopodal setae visible through the integument.

A convex midline is exposed in a few specimens. There appear to be four shallow furrows indicating segmentation (Figs. 57A, G). Specimen CN 115318 has corresponding furrows on the ventral side (Fig. 59A). This indicates the presence of five segments in the tail. Exopod setae are present under the tail shield (Fig. 59A), indicating the presence of biramous legs also in this part of the animal.

The gut is seen as a thin dark band without mud filling in a large specimen, which is not illustrated herein.

Discussion. – The dorsal reconstruction (Fig. 60) is based on two specimens that are bilaterally symmetrical and broadly oval in outline, with parallel compaction (Figs. 57A, G). The original outline seems to have been broadly rather than narrowly oval. In one of them in particular (Fig. 57G), the arched posterior margins of the head shield and anterior thoracic tergites indicate that these elements were tilted a little backward before compression. The arching thus reveals something of the original convexity. There was no axial lobe and furrows but a rather narrow 'hogback', from which the lateral sides tilted down like the roof of a house. The convexity apparently decreased progressively from the head to the tail shield. The presence of a slight overlap between tergites only in the axial region must have made the exoskeleton almost inflexible.

Preparation of specimen CN 115318 did not reveal either hypostome or antennae (Fig. 59A). The anterior-most leg seen may belong to the right side. If so, it is turned over to the left and placed close to the rostral plate.

An endite appears to be present proximal to the basis in the 2nd left thoracic limb (en5 in Fig. 59A). It is similar to the basis in its spinosity. The 4th left thoracic limb (en7 in Fig. 59A) shows a similar endite.

Kuamaia muricata n. sp.

Figs. 61–62

Name. – Latin *muricatus*, with spines like those of the gastropod *Murex*.

Holotype. – CN 115401 from Maotianshan, level M2. This is the only known specimen.

Diagnosis. – Exoskeleton elongately oval in outline, with eight thoracic tergites; tail shield with terminal and three pairs of marginal spines; rostral plate large and convex, not protruding beyond anterior margin of head shield; endopod strong, angular in cross section; both main and distal elements of exopod wide and large, the distal element bearing fine bristles distally.

Description. – (a) General characteristics: As preserved, the exoskeleton is flattened, but it is symmetric and lacks wrinkles or folds. The length is about 14.8 mm, the maximum width is 8.7 mm.

As the split exposing the specimen is along a plane that is oblique to the bedding of the mudstone, the lower part of the specimen is incomplete (Figs. 61B, 62B), whereas the upper part is almost complete (Figs. 61A, 62A). The description of the head shield is based on the upper part.

(b) Head: The head shield is subelliptical in outline, with a smoothly rounded anterior margin (Figs. 61A, 62A). There are short genal (posterolateral) angles. The rostral plate does not seem to protrude beyond the anterior margin of the tergum (Fig. 62A).

A pair of oval pits occur anteriorly on the head shield; the left one is irregular in shape, probably through oblique burial and subsequent deformation (Fig. 62A). The two pits should correspond to convexities on the lower part. They occur where eyes are sited in *Kuamaia lata* and in all likelyhood represent compound eyes.

The head has no defined axial region. The surface is smooth except for impressions of ventral structures caused by appendages.

(c) Body: The body has eight thoracic tergites and a triangular tail shield. The first three tergites are of equal length, the five succeeding ones progressively shorter; the second tergite is broadest. The lateral marginal spines increase in length from the head to the third thoracic tergite, from where they are of constant length.

The thoracic tergites weakly overlap only medially; there is no overlap in the pleural region (Fig. 61A–B).

The triangular tail shield (Figs. 61B, 62B) is both narrower and shorter (excluding the terminal spine) than the head shield. The lateral spines are equal in length to those of the preceding thoracic tergites. The terminal spine is a little longer than the lateral spines. An outwardly and backwardly directed shallow furrow occurs on the left side of the tail shield.

(d) Ventral side: The anterior margin of the rostral plate joins the margin of the head shield (rather than projects in front of the tergum, as in *Kuamaia lata*), if our interpretation of the rostral plate as having a ventral doublure is correct. Behind the rostral plate is the presumed, oval-shaped hypostome, concave in the upper part, and delimited anteriorly by the rostral plate and laterally and posteriorly by a faint ridge (Figs. 61A, 62A). The antennae originate from the sides of the presumed hypostome and extend beyond the lateral margin of the head shield, one being directed forwards–outwards, the other extending more to the side. The antennal annuli are impressed on the head shield, each being about equal in length and width (Figs. 61A, 62A). No setae are seen on the antennae.

A longitudinal furrow in the head shield represents an impression of the alimentary canal (Figs. 61A, 62A). It begins just behind the presumed hypostome. Three transverse furrows representing limb impressions extend laterally from the longitudinal furrow (Figs. 61A, 62A). The first two furrows are close to one another and appear to merge proximally, which indicates that they may represent the two branches of a single limb. Behind the third impression, which may locate the 2nd leg, there is a pair of presumed apodemal pits (or attachment surfaces) close

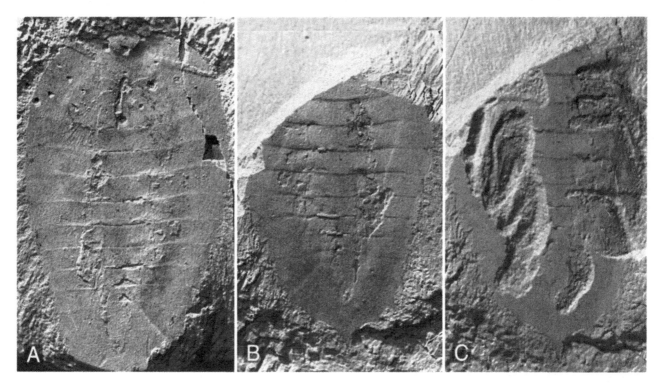

Fig. 61. Kuamaia muricata n.sp. Holotype, CN 115401, from Maotianshan, level M2 . □A. Upper part, ×6.3. □B. Lower part before preparation, ×6.3. □C. Lower part after preparation, ×6.9.

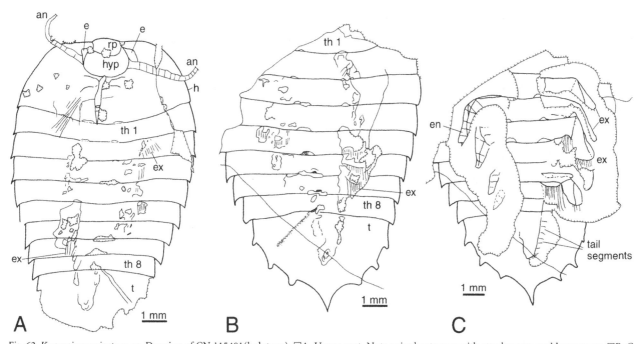

Fig. 62. Kuamaia muricata n.sp. Drawing of CN 115401(holotype). □A. Upper part. Note paired antennae with attachments, and hypostome. □B, C. Lower part before and after preparation. The latter shows endopods on left side, exopods to the right. For abbreviations, see Fig. 9.

to the rear of the head shield, indicating a third pair of legs. Exopodal setae impressed on the head shield appear to belong to the 1st left exopod. Paired small semicircular structures on the thoracic tergites may represent apodemal pits, or anyway relate to the limbs, which shows that

there is one pair of legs for each thoracic tergite. The structures are close to the posterior margin of each tergite.

By preparation of the lower part specimen, some limbs have been exposed, endopods shown on the left, exopods on the right (Figs. 61C, 62C). The endopod is strong, curved outward and downward; five podomeres and one additional terminal element can be discerned. One or two ridges extend along the wide endopod, indicating that it was perhaps triangular in cross section. Both main and distal elements of the exopod are wide. The setae are long and thin. The distal exopod element is directed backwards–outwards. Its surface is marked by a ridge, and it has fine bristles distally. A shallow furrow extending from the last left thoracic supposed apodeme (Figs. 61B, 62A) probably represents a limb of the last thoracic segment.

There are about 8 body segments along the right side of the tail axis (Figs. 61C, 62C).

Discussion. – The specimen of *Kuamaia muricata* is smaller than specimens of *K. lata*, but we do not consider it to be a young individual of the latter. The rostral plate is very different in position; it has one thoracic segment more than *K. lata* and an additional pair of spines on the tail shield. Particularly pertinent is the fact that the number of thoracic segments may increase, but not decrease, with size.

The shortness of the tail segments is notable. If the segments carried appendages, these must have been considerably weaker than in the thorax and probably without much function in locomotion, because they would have been too crowded to work efficiently.

The specimen is dorsoventrally compressed, bilaterally symmetrical, elongate–oval in dorsal outline, and has a non-wrinkled surface. Such features are commonly encountered in Burgess Shale and Chengjiang specimens and cannot be taken to indicate a very low relief of the original exoskeleton. The degree of wrinkling presumably to some degree depends on the original stiffness. *K. muricata* resembles some of the *K. lata* specimens in the almost straight anterior margin of the head shield and the post-compressional convex posterior margin of the head shield (cf. Figs. 57A, F, with 61A, 62A). The original convexity of the exoskeleton in *K. muricata* appears to be similar to that of *K. lata*.

Genus *Rhombicalvaria* Hou, 1987

Type species. – *Rhombicalvaria acantha* Hou, 1987.

Rhombicalvaria acantha Hou, 1987

Synonymy. – □1987b *Rhombicalvaria acantha* gen. et sp.nov. – Hou, p. 284, Pl. 3:3–4, Text-fig. 5. □1991 *Rhombicalvaria acantha* Hou – Hou & Bergström, p. 183. □1991 *Rhombicalvaria acantha* Hou – Delle Cave & Simonetta 1991, p. 201, Fig. 6D.

Holotype. – CN100135 (Hou 1989b, Pl. 3:3).

Distribution. – Both the holotype and the paratype (CN 100136, Hou 1987b, Pl. 3:4) are from Maotianshan, level M2.

Discussion. – We have no additional information on this species. It seems to correspond well with other species of the family, being characterized basically by its longer spines and by the specific number of spines on the tail shield.

Family Skioldiidae n. fam.

Diagnosis. – Helmetiids with weakly defined axial region; segmental boundaries fused in tergum, poorly developed laterally, which means that head, thorax and pygidium are not fully differentiated; margin serrated.

Genus included. – *Skioldia* n. gen.

Genus *Skioldia* n. gen.

Name. – From old Norse *skiold*, shield, referring to the shape.

Type species. – *Skioldia aldna* n. gen. et sp.

Diagnosis. – Broadly oval tergum with nine or ten thoracic semitergites and a triangular tail shield; head and tail portions have shorter furrows; an axial region is vaguely delimited; one pair of eyes are situated fairly close to the rostral plate; margin entire, but with small pointed spinules.

Discussion. – *Skioldia* resembles *Saperion* but is much broader and shorter, with some 12–13 visible segments, as compared to about 20 in *Saperion*.

Skioldia aldna n. gen. et sp.

Figs. 63–65

Name. – Old Norse *aldna*, old.

Holotype. – CN 115402, a complete specimen including upper and lower parts.

Other specimen. – CN 115403, a paratype specimen including upper and lower parts.

Diagnosis. – As for the genus.

Distribution. – Both specimens are from Maotianshan, level M2.

Description. – (a) General characteristics: The entire dorsum forms a broad, smooth single shield, about 68 mm long and 58 mm wide in the holotype and 25 mm long and about 20 mm wide in the paratype. In the paratype

Fig. 63. Skioldia aldna n.gen. et sp. Holotype, CN 115402, from Mao-tianshan, level M2, lower part, showing a depression in the head caused by the antenna. ×1.2.

the tergum is compressed to a white film (Fig. 64A–B). In the holotype the right half of the tergum is smooth, whereas the left half is grainy owing to the coarse matrix (Fig. 63). The margin of the whole tergum is serrated with small, pointed spines (Figs. 63, 65A). There is possibly a vaguely defined axis, up to about 17 mm wide (Fig. 63). Alternatively, this region is only the result of the pleura being deformed to to the horizontal plane.

In the middle part of the body, which may be compared with a thorax, segment boundaries are mostly distinct and step-like for most of their extent. The steps disappear towards the margin of the shield, which means that the 9–10 successive 'semi-tergites' meet edge to edge, and the boundaries are virtually lost at the margin (Figs. 63–65). A tiny indention at the margin marks the place of the segment boundary. Each 'semi-tergite' widens laterally, from around 3 mm at the axis to 5 mm at the shield margin. The 'head' and 'tail' parts of the shield have at least three and two segment boundaries, respectively, which are distinct in the middle but disappear well before they reach the shield margin (Figs. 63, 65A).

The margin is almost smooth in the poorly defined head and thorax, with only quite small, regularly

arranged, pointed spines. In the 'tail' the spinosity becomes much more marked and irregular, with the spines arranged in groups delimited by larger, more rounded spines and adjoining notches, which may mark segment boundaries. The poorly defined tail has spines 0.1–0.8 mm long and is terminated by a slightly longer spine (Fig. 65).

The only structural differentiation is seen in the head. A pair of rounded structures fairly close to the anterior margin no doubt represent compound eyes. The anterior margin is almost straight, with shallow notches in front of the eyes. The narrow strip between the notches slopes forwards. An axial portion, which appears raised, has its origin a short distance behind the eyes. A furrow, extending from the anterolateral corner of the axis along the posterior side of the eye and posterolaterally over the cheek, represents the collapsed right antenna which has about 35–40 annuli visible. No other appendages are seen in the specimen.

On the right side of the paratype, there is a curved blade-shape structure (Figs. 64A, 65B). Several similar objects have been collected from the fauna, showing articulating structures at the anterior and posterior portions of the curved part. It appears to represent a thoracic tergite of an arthropod, possibly the last tergite of *Acanthomeridion serratum* (cf. Fig. 36).

(b) 'Head': The anterior, poorly segmented part of the shield is comparatively broad and short, and its anterior margin is somewhat retracted with a large, central, rostral plate.

The rostral plate appears to protrude notably beyond the anterior margin of the head shield (Fig. 64A). A linear structure extends laterally from the posterior margin of the rostral plate to the anterolateral margins of the shield (Figs. 63–65).

In the axial part of the head, there are three more or less distinct segment boundaries, which disappear before they reach the shield margin (Figs. 63A, 65A).

(c) 'Body': The holotype (Figs. 63, 65A) shows nine thoracic semitergites in a 'thorax', the paratype seems to have ten. Toward the margin of the shield, the anterior-most semitergites bend slightly forward, the posterior ones backward. All semitergites are shorter axially than at the shield margin, the lengths being about 3 and 5 mm, respectively. The third and fourth tergites form the widest part of the animal.

The margin has many small, pointed spines, which become much more marked and irregular posteriorly. The spines are 0.1–0.8 mm long and arranged in groups delimited by larger spines and adjoining notches that mark segment boundaries (Fig. 65).

The 'tail' shows at least two segment boundaries, which are visible only in the axial region (Figs. 63, 65A). The peripheral spines of the tail are like those of the thoracic tergites in arrangement and size, with the spines in groups

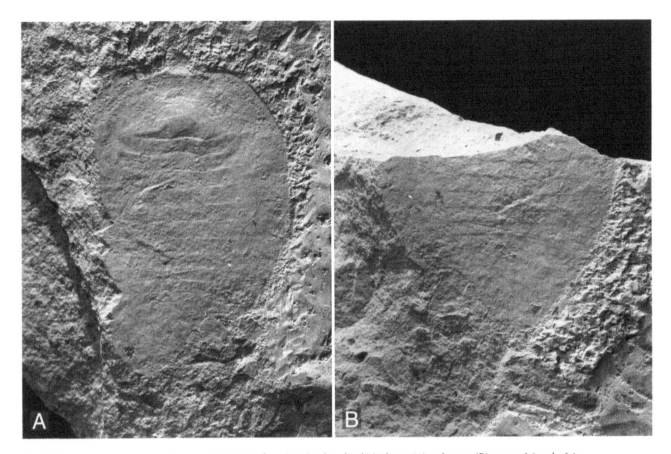

Fig. 64. Skioldia aldna n.gen. et sp. Paratype, CN 115403, from Maotianshan, level M2, lower (A) and upper (B) parts, ×2.1 and ×3.1.

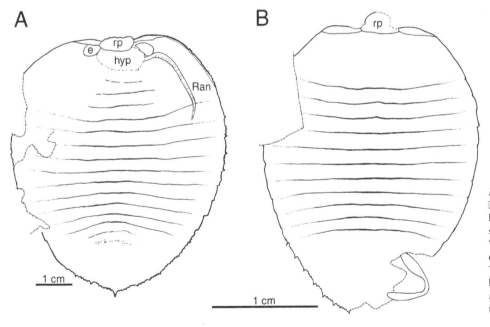

Fig. 65. Skioldia aldna n.gen. et sp. □A. Drawing of lower part of the holotype, CN 115402. Note the short segment boundaries in the 'head' and 'tail'. □B. Composite drawing of paratype CN 115403. The curved thoracic tergite at the bottom right belongs to *Acanthomeridion serratum*. For abbreviations, see Fig. 9.

delimited by large spines, which may mark segment boundaries (Fig. 65). There appear to be 5–6 pairs of larger spines, thus probably indicating that the tail consists of 5–6 segments. The terminal tail spine 1.5 mm long.

(d) Ventral side: In the body part considered as the head, a very narrow convex band along the margin in the lower part indicates the presence of a narrow doublure (Fig. 63).

The rostral plate seems to be succeeded by a large elliptical structure (possible hypostome), from the side of which the right antenna appears to have its origin (Figs. 63, 65A). The right antenna is mirrored by a furrow on the shield (Fig. 63), extending from the side of the possible hypostome along the posterior side of the eye and posterolaterally over the cheek. Many annuli are visible; there are a total number of about 35–40, the length of each being about equal to or slightly shorter than its width. No other appendages are seen in the holotype. The intestine has not been seen.

Discussion. – *Skioldia aldna* is very similar to *Kuamaia lata* in the outline of the tergum and features of the head shield, such as the eye position close to the rostral plate, dorsal facial sutures, and a vaguely defined axial region (cf. Figs. 57–59), but differs in having a serrated shield margin. As in *Kuamaia lata*, there seems to be a separate plate just behind the rostral plate (Fig. 63), which resembles the ovate rostral plate in outline and is delimited by a shallow furrow which touches the eyes. The plate may represent a hypostome with a position corresponding to that of *Kuamaia*.

The width of *Skioldia aldna* is 86% of the length, and hence it is much broader than *Saperion glumaceum* (48%) and the holotype of *Tegopelte gigas* (49%).

Family Saperiidae n.fam.

Diagnosis. – Helmetiids lacking distinct boundaries between cephalon, thorax and pygidium, lacking axis, and having an entire margin.

Genus included. – *Saperion* Hou, Ramsköld & Bergström, 1991.

Genus *Saperion* Hou, Ramsköld & Bergström, 1991

Type species. – *Saperion glumaceum* Hou, Ramsköld & Bergström, 1991.

Emended diagnosis. – Narrowly oval tergum with about thirteen thoracic segments, followed by shorter segments which may be counted with the tail; head with rostral plate and devoid of furrows; tail portion elliptical in outline, with shorter furrows; margin entire, smooth.

Discussion. – *Saperion* is similar to *Kuamaia, Rhombicalvaria* and *Skioldia* in having an ovate rostral plate, a dorsal facial suture and in the nature of the segmental boundaries, but differs from those genera in its narrowly oval outline in dorsal view and its smooth peripheral margin. Hou

et al. (1991) considered. that the thorax embraced the entire part of the body with signs of segmentation. After considering segmentation in *Kuamaia* and *Skioldia*, we now think that the posterior part with shorter segments and much shorter boundary lines should be considered as part of the tail. *Saperion* is similar to *Tegopelte* in outline, but the latter apparently has no signs of segmentation in the tergum and no rostral plate in the head, and may have seleniform compound eyes. No eyes are seen in the single specimen of *Saperion glumaceum*, but their absence may be a factor of preservation, or they may have had a ventral position.

Saperion glumaceum Hou, Ramsköld & Bergström, 1991

Figs. 66–67

Synonymy. – □1991 *Saperion glumaceum* sp. n.– Hou *et al.*, pp. 401–402, Fig. 3A. □1996 *Saperion glumaceum* Hou *et al.* – Ramsköld *et al.*, , Fig. 1A.

Holotype. – CN 115289, a complete specimen with upper and lower parts from the eastern side of Jianbaobaoshan, about 300 m west of Dapotou village, Chengjiang, level Dj1.

Other material. – None.

Description.. – (a) General characteristics: The dorsum forms a smooth, single shield. The exoskeleton is thin and preserved as a whitish film. Compression has caused irregular folding and wrinkling, especially anteriorly (Fig. 66). The margin is even, without any spines or notches marking segment boundaries. The character of the segment boundaries makes it possible to distinguish 'head', 'thorax' and 'pygidium' within the shield. The tergum is narrowly oval in outline, 23.1 mm long and 11.2 mm wide. A much larger specimen, 151 mm long, is illustrated by Ramsköld *et al.* (1996, Fig. 1A). The greatest width is around the second segment of the poorly defined thorax. From there the tergum narrows rapidly forwards and gradually backwards. The gut has not been seen.

(b) 'Head': An anterior portion, 4.6 mm long, is devoid of segmental furrows and constitutes the 'head'. It is quite similar to that of *Kuamaia lata* and *Skioldia aldna* in outline. The ovate rostral plate protrudes beyond the otherwise almost straight anterior margin. The suture or line delimiting the pararostral plate posteriorly is similar to that in *Kuamaia lata* and *Skioldia aldna*. Paired eyes and segmental furrows are not seen, possibly because of flattening and distortion of the specimen. Likewise, the absence of features indicating distinction of an axial area may not have characterized the living animal.

(c) 'Thorax': The thorax has 13 (rather than 17 in Hou *et al.*, 1991) segments delimited by furrows that fade out

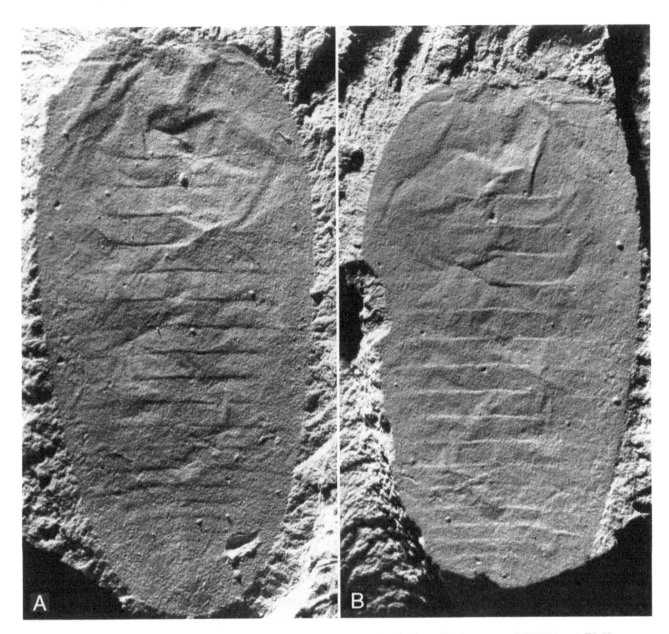

Fig. 66. Saperion glumaceum Hou *et al.*, 1991. Holotype, CN115289, from Jianbaobaoshan, level Dj1. □A. Lower part, CN 115289a, ×6. □B. Upper part, CN 115289b, ×6.

towards the margin of the shield. The first six segments are of equal length, while the succeeding nine successively decrease in length. In the central part of the shield, each segment boundary marks the position of a step where the successive segment steps down below its neighbour in front.

(d) 'Tail': The tail is elliptical in outline and has at least six segments, which are shorter and less distinctly delimited than in the thorax (Figs. 66A, 67). The segmental furrows end well inside the margin of the shield. The midline is convex in the tail tagma. This may indicate a convex axis or may be taken as an indication of a more hogback-shaped original convexity of the entire dorsum.

Discussion. – The discussion (see above) on the flexibility of the shield in *Skioldia aldna* also applies to *Saperion glumaceum*. The tergum is decidedly more wrinkled by compaction than is the case in species of *Kuamia*. One reason may be that the specimens are quite small and delicate, whereas *Kuamaia lata* is larger and sturdier. It is also possible that *S. glumaceum* had a stronger convexity. The wrinkling is additional to the segmental boundaries, which are very regular both in direction and spacing and in the little step down from one segment to the succeeding one. In this species, therefore, they are not preservational artefacts as suggested by Ramsköld *et al.* (1996). By contrast, the transverse lines described from *Naraoia* (Ram-

and lacking mineralization; dorsal eyes in certain species only; feeding through mud ingestion.

Orders. – Xandarellida Chen *et al.*, 1996, and Sinoburiida n.ord.

Order Xandarellida Chen, Ramsköld, Edgecombe & Zhou *in* Chen *et al.* 1996

Diagnosis. – Petalopleurans of trilobite-like habitus, with large head and tapering trunk, the latter posteriorly with tergites that may cover more than one segment. Tergites free and overlapping. Endopods multisegmented, particularly in head.

Families. – Xandarellidae n.fam. and Almeniidae n.fam.

Family Xandarellidae n.fam.

Diagnosis. – Xandarellids with dorsal compound eyes and an incompletely fused head tergite.

Genus included. – *Xandarella* Hou, Ramsköld & Bergström, 1991; *Cindarella* Chen, Ramsköld, Edgecombe & Zhou *in* Chen *et al.* 1996.

Genus *Xandarella* Hou, Ramsköld & Bergström, 1991

Type species. – *Xandarella spectaculum* Hou, Ramsköld & Bergström, 1991.

Diagnosis. – The unmineralized exoskeleton is divided into two large head tergites, one small tergite without pleura, seven thoracic tergites, and four abdominal tergites. One pair of antennae and six pairs of biramous legs under the posterior head tergite. One pair of biramous legs under the small tergite and under each of thoracic tergites. Two pairs of biramous legs under the 1st abdominal tergite, four pairs under the 2nd, five pairs under the 3rd, and at least 12 pairs under the posterior-most tergite, which has a terminal spine. Structure of exopods complicated; two rows of setae on flanges of the 3rd podomere, one row of bristles and one row of setae on the distal podomere. Endopods multisegmented, but exact number of podomeres uncertain. Each cephalic endopod may have 12 podomeres and a terminal element, each thoracic one may have 11 podomeres and a terminal element.

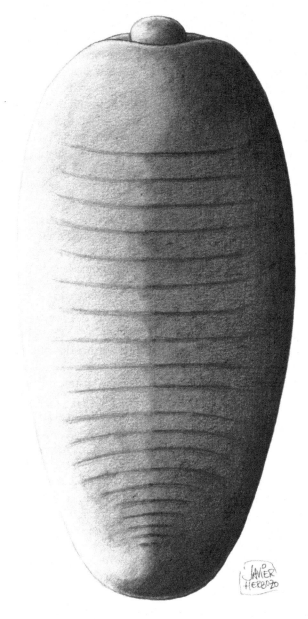

Fig. 67. Saperion glumaceum Hou et al., 1991. Reconstruction in dorsal view.

sköld *et al.* 1996, Fig. 1B) and *Tegopelte* (Whittington 1985a) are irregularly spaced, and their inclination to the longitudinal axis of the animal shifts from one line to the next. We agree that in these two species, the lines are preservational artefacts.

Subclass Petalopleura n.subcl.

Name. – Greek *petalos*, broad, flat, outspread, and *pleura*, side.

Diagnosis. – Artiopodan arthropods with well-developed pleural folds, distinct overlap between adjoining tergites,

Xandarella spectaculum Hou, Ramsköld & Bergström, 1991

Figs. 68–74

Synonymy. – □1991 *Xandarella spectaculum* sp. n.– Hou, Ramsköld & Bergström, pp. 402–403, Fig. 3B.

Holotype. – CN 115285, lower part only, a complete exoskeleton with antennae. The right exopods under the head shield have been exposed by the removal of parts of the tergum. The holotype is from level M3 at Maotianshan.

Other specimens. – Two additional specimens. In CN 115286, upper and lower parts of an almost complete individual (mentioned by Hou *et al.* 1991, p. 402), preparation has revealed details of well-preserved legs. CN 115404 was divided into two parts when the rock was split. The lower part comprises the posterior portion of the animal, the upper part the anterior portion.

Distribution. – Maotianshan, level M3, and Jianbaobaoshan, level Dj2.

Description.. – (a) General characteristics: The animal is oval in dorsal aspect, widest at the genal angles and first thoracic tergite. In the holotype and CN 115286 (Figs. 68D, 69, 70), the exoskeleton is dorsoventrally flattened but with some convexity, with both sides slightly sloping from the axial part. In the third specimen (CN 115404; Figs. 68A–B, 71F–G), the exoskeleton preserves more convexity. The tergum is not visibly distorted, indicating that the animal was of moderate convexity even in life. The surface is smooth. In the axial region, impressions indicate the presence of ventral appendage structures. The overlap between tergites is seen as dark bands. Reddish brown spots are often scattered over the surface. The colour may be caused by ferric oxide stain.

All three specimens are of similar size. The holotype has a complete outline in dorsal view and is 51 mm long and 29 wide.

(b) Head: The head shield is semi-elliptical, with slightly acute genal angles. The posterior margin is gently concave laterally and has a strong median embayment to encompass the small successive tergite (Figs. 68D–E, 70, 71A–B, D, G).

Paired compound eyes are placed laterally a little more than half-way from the midline to the lateral margin and about half-way between the anterior and posterior ends of the head. The head appears to be composed of two large tergites, which overlap laterally of the eyes. The boundary appears to be present also between the eyes, where it bends forwards and is mostly seen only as two weak lines (Figs. 68D–E, 71B, D). It is possible that there is a partial fusion between the tergites in the area between the eyes. Each eye is surrounded by a coloured zone, indistinguishable from that of the overlap. This probably means that

the eye is situated on a discrete little round tergite. The visual surface of the eye is kidney-shaped, narrower anteriorly than posteriorly, raised above the surrounding surface, and appears to consist of hexagonal lenses (Fig. 68E). The number of lenses is estimated to be over 1000.

(c) Body: At the transition between head and thorax, in the posterior embayment of the head shield, there is (in CN 115285 and 115286) a small tergite without pleura. The overlap between the small tergite and neighbouring tergites is normal (Figs. 68D–E, 70, 71A, B, D, G). It is a matter of convenience whether this tergite is counted with the head or with the body.

Behind the small tergites, there are another seven thoracic tergites and four abdominal tergites. The pleural spines are posteriorly directed and progressively longer backwards. The thoracic tergites are of equal length; the abdominal tergites are successively longer backwards. The last tergite has a terminal spine (Figs. 68B, 70, 71A, F). In the holotype, the spine base is marked by a concavity at about the middle of the tergite (Figs. 70, 71A). The spine is triangular in cross section, with a distinct dorsal keel (Fig. 68B).

The overlap between the tergites is distinct, being marked by a dark band (Figs. 68D–E, 69, 70, cf. 71). The first body tergite is overlapped axially by the small tergite, laterally by the head shield (Figs. 71A, B, D, G). The amount of overlap is particularly small in the axial area. Axially, the underlapping tergite forms an 'articulating half-ring', which appears as a narrow raised band in the anterior half of the overlap (Figs. 68D, E). In each tergite a shallow pleural furrow extends from the pleural spine to the axial area and forward to meet the posterior margin of the preceding tergite (Figs. 68B, D, E, 70, 71A, B, D, F, 73).

(d) Ventral side: The antennae are well preserved in the holotype and in CN 115286 (Figs. 68D, E, 69A–B, 70, 71A–B, D). They are clearly seen extending beyond the anterior margin of the head. The proximal portion of the antennae is indicated by compressed furrows on the anterior head tergite (Figs. 68D, 70). Even here, segmentation is distinct (Figs. 68D, E, 71B, D). In CN 115286, the right antenna is completely preserved in the upper-part specimen and consists of 46 annuli, of which about 9 proximal annuli seem to be of equal length and width and the others longer than wide (Fig. 71D, E). The antennae clearly attach at the sides of a subelliptical structure presumed to be a hypostome, which is not separate from the margin but is elevated over its surroundings (Fig. 71A, B, D, G). In specimen CN 115404 (Figs. 68A, 71G), the supposed hypostome is slightly convex, indicating some degree of sclerotization that could withstand post-mortem compression. The mouth is at the posterior tip of the hypostome (Fig. 74), as indicated by a small round structure (Figs. 68A, 71G, 72). The gut is traced from the middle of the hypostome, where it forms an anteriorly directed loop (Figs. 71A, B, D, G).

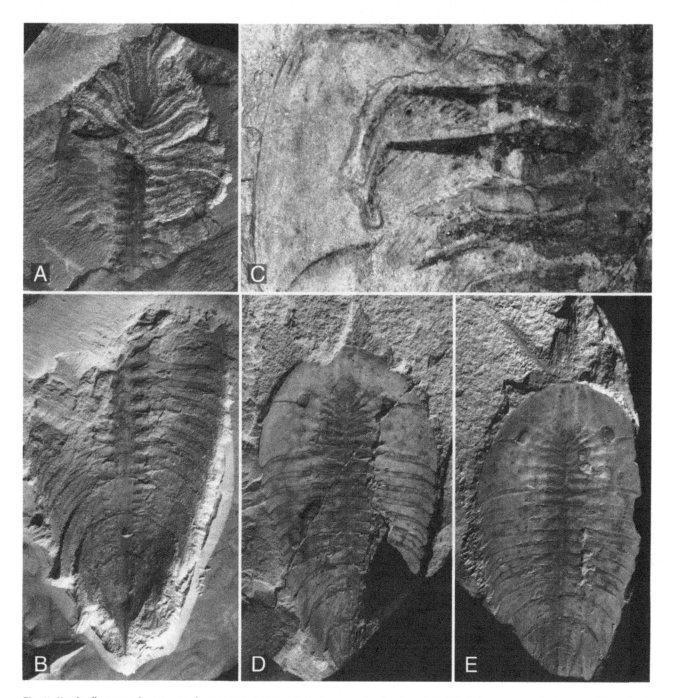

Fig. 68. Xandarella spectaculum Hou *et al.*, 1991. □A, B. CN 115404, from Maotianshan, level M2. □A. Upper part, anterior portion of the animal, exposed appendages in ventral view, cf. Fig. 71G, ×1.7. □B. Lower part, posterior portion of the animal, showing a moderate convexity of the exoskeleton, cf. Fig. 71F, ×2.1. □C–E. CN 115286, from Jianbaobaoshan, level Dj2. □C. Left appendages on the head, after preparation. Note the anterior axis and posterior flange of exopods and the two proximal podomeres of 6th left endopod. ×7.4. □D. Lower part before preparation, showing two weak lines between eye pair and intestine, cf. Fig. 71B, ×1.9. □E. Upper part, a complete antenna preserved, cf. Fig. 71D, ×1.9.

Behind the hypostome are six pairs of biramous legs under the head shield (Figs. 68D–E, 71B, D). This is one more pair than previously reported (Hou *et al.* 1991, p. 402). The first pair is located laterad or posterolaterad of the mouth. The small tergite covers one pair of biramous legs. Likewise, each of the successive seven thoracic tergites covers one pair of biramous legs. In contrast, the abdominal tergites cover two (first tergite), four (second tergite), and five (third tergite) pairs of legs (Figs. 68B, D, E, 70, 71A, B, D, F). The number of appendages under the last abdominal tergite is uncertain, but impressions of six or seven leg pairs are seen in the anterior part (holotype and CN 115404; Figs. 68B, 70, 71A, F). There appear to be additional legs further back. In CN 115286, the three-

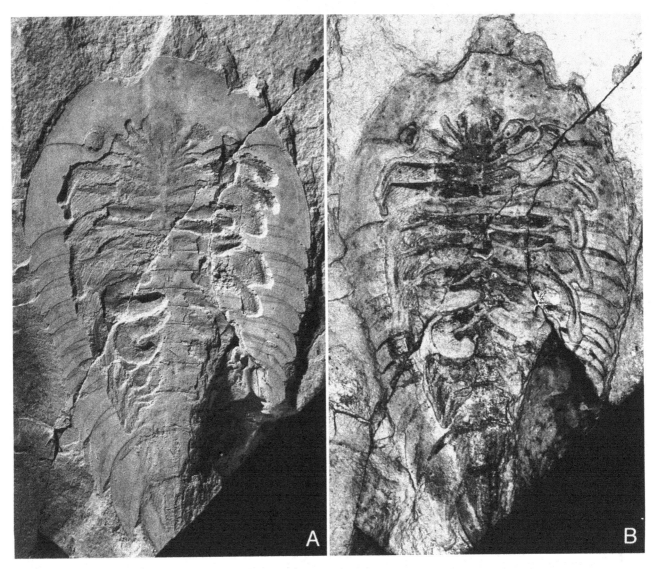

Fig. 69. Xandarella spectaculum Hou *et al.*, 1991. CN 115286a, lower part, prepared to show appendages; cf. Fig. 72. □A. Covered with ammonium chloride, ×2.8. □B. Photographed with panchromatic film and non-directional light, ×2.8.

dimensionally preserved straight gut appears to extend to the base of the terminal spine (Figs. 68E, 71D), and 12 or 13 imprints of legs occur at the sides of the gut (Figs. 68D, 71B). It seems plausible that there are at least 12 leg pairs corresponding to the last abdominal tergite. Impressions of exopod setae are clearly shown in the pleural area (Figs. 68D, E, 71B, D, 72), indicating a biramous nature of the abdominal appendages.

Preparation of CN 115286 has revealed the detailed structure of the appendages (Figs. 69A, B, 71C, 72). The appendages are not pendent but extend laterally. The exopods are similar throughout the body.

On the right side of the head of the holotype, successive exopods are seen to overlap each other backwards even in the proximal portions (Figs. 70, 71A). The proximal portion of the exopod is composed of two segments and is

succeeded by a long element with two rows of setae and a distal element with at least one row of bristles and one row of setae. The proximal segments can be clearly seen on the right anterior exopods (especially on the 6th and 7th right exopods in the holotype), in which the proximal exopod segments show a longitudinal line that may somehow be an impression of the endopod, although hardly of the posterior edge (Fig. 71A). The distal portion of the second segment forms an expansion to which the long third element is attached.

As seen from above, there are three or four setae on the posterior margin of the expanded portion of the second podomere, visible on the 6th and 7th right exopods in the holotype (Figs. 70, 71A). In dorsal view, a straight furrow divides the true axis in front from a flange behind (cf. below). The anterior axis is narrow and of even width.

The posterior flange is wide proximally but tapers gradually distally; 13–15 bladelike setae are each attached to its posterior margin by a clear articulation (thus verifying its setal nature). The line of articulations appears as a furrow between the flat flange and the setae, which are rotated slightly so that the flat surface tilts towards the midline of the body. As revealed in the exopods of CN 115286 (Figs. 69, 72), the proximal setae are longer than distal setae. The setae are shorter than the distal element of the shaft. Along the entire length of each seta, the dorsal rim is swollen and has a row of 35–45 bristles. These appear to be narrower at their bases, which may indicate a basal articulation. The arrangement and shape of the bristles are seen on the right 5th and 8th exopods in CN 115286 (Fig. 71C).

The row of setae was removed by preparation to reveal a second row tilted in the same direction as those of the upper row (Fig. 71C). Each of the two rows appears to attach to a flange of the axis; presumably the longitudinal line seen from above demarcates the boundary between the anterior axis and the posterodorsal flange. Where the lower surface (or rather its mould) of the axis has been made visible by preparation close to where the upper side is seen, it is evident that the two surfaces are flat and diverge from the axis. The axis thus is triangular in cross section. The two flanges extend as prolongations of the upper (dorsal) and lower (ventral) surfaces of the axis, and the distinction is only marked by a distinct line. Not only the axis, but also the two flanges appear to attach proximally to the swollen distal portion of the 2nd exopod segment. The flanges appear to be immovably attached to the axis. About 13–15 setae are attached to the posterior margin of each flange, again in the same two planes as the surfaces of the axis and flanges. Three to four additional setae attach to the posterior margin of the swollen portion of the second proximal segment (Figs. 70, 71A). The upper row extends above the exopod of the successive segment, whereas the lower row apparently dips in under this exopod. This means that the distal element of the exopod is partly embraced by the setae of the proximal element.

The distal element of the axis is long and rod-like, about $\frac{2}{3}$ the length of the penultimate element, attached to this element by an articulation which is visible as a transverse furrow. The distal element has somewhat different postures in different appendages. This indicates that the articulation was flexible (Figs. 69A–B, 72). A line of about 40 long bristles (or hairs) are attached to a dorsal longitudinal ridge, as seen on the distal elements of the 2nd, 4th and 5th right exopods. Further excavation revealed a row of blade-like setae attached to the posterior edge of the distal element (Figs. 69A–B, 71C). The setae appear to be rotated in the same direction as those of the penultimate podomere. The underside of the axis has not been seen. The setae of the penultimate element extend backward

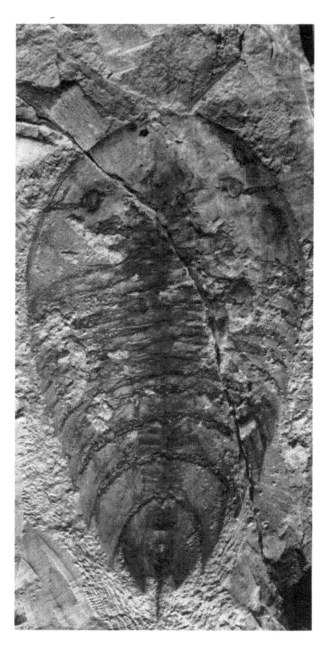

Fig. 70. Xandarella spectaculum Hou *et al.*, 1991. Holotype, CN 115285, from Maotianshan, level M3, lower part, showing the proximal characteristics of the successive exopods on the right side of the head; cf. Fig. 71A. ×2.6.

and outward, where they may overlap part of the distal element. In life, the bristles and setae of the distal element appear to have extended partly between the two sets of setae of the penultimate element.

Preparation of CN 115286 has revealed the right endopods of the head (Figs. 69, 72). Some $\frac{3}{5}$ of the appendage extends anterolaterally, whereas the distal part is folded back on the proximal part. The entire endopod is slender and so long that it appears to be able to reach to or near to the lateral margin of the head shield when extended. The

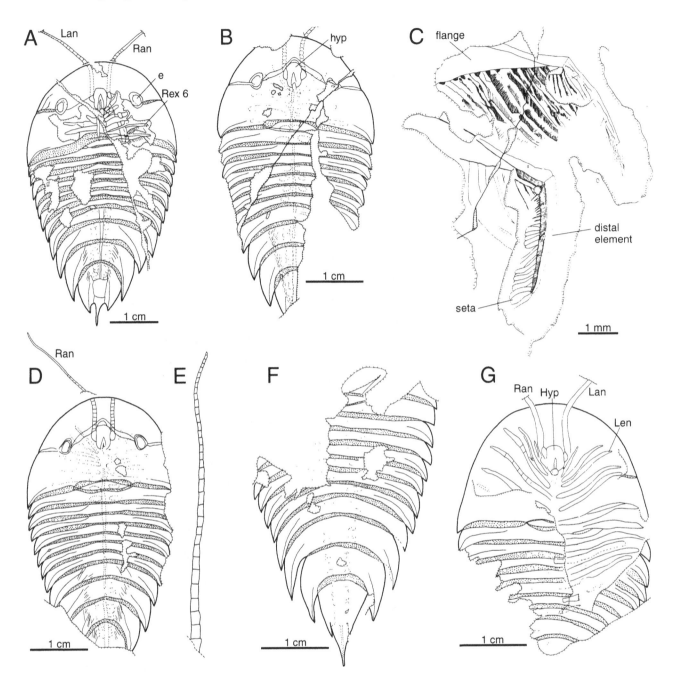

Fig. 71. Xandarella spectaculum Hou et al., 1991. ☐A. Drawing of CN 115285, holotype, lower part. Note the proximal segments of the exopods shown on the right side of the head. ☐B. Drawing of CN 115286a, lower part, before preparation. ☐C. Detail of distal portion of the 5th and 6th right exopods in CN 115286a, after preparation. ☐D. Drawing of CN 115286b, upper part. ☐E. Detail of right antenna in CN 115286b. F, drawing of CN 115404a, lower part. ☐G. Drawing of CN 115404b, upper part, after preparation. For abbreviations, see Fig. 9.

boundaries between the podomeres are only faintly visible. There are about 12 podomeres, which seem to be much longer than wide, and a terminal element in each endopod. On the 6th left endopod, which is three-dimensionally preserved and partly exposed by preparation, a short ventral spine is shown on what appears to be the 2nd podomere, but not on the 1st podomere. Short ventral spines are also shown on distal podomeres of the 3rd right

endopod. This suggests that, in the head appendages, there may have been a short ventral spine on each podomere, except for the 1st podomere. The terminal element of the 3rd right endopod has a ventral spine. On the 3rd right endopod, dorsal spines (or possibly setae, cf. below) occur on the 3rd and about the 6th podomeres, where they are held tight to the dorsal margin of the endopod. A dorsal spine on the 1st podomere of the 6th left endopod is bent out at 40° from the endopod. This difference in posture may indicate that the spine-like structures have an articulation at their base.

Occasionally a pair of small, ventral plates are seen between opposing endopod bases, as between the 5th, 6th and 9th pairs of endopods (Figs. 69A, B, 72). These plates appear to carry three small teeth at the mediad side. The teeth of one plate interdigitate with those of the opposing plate. The plates may be regarded as sternites.

The endopods of the thorax appear to be much more robust than those of the head. In the compressed fossil, they are folded backwards rather than forwards as in the head. They are strongly curved, so that the terminal podomere is located near the midline of the animal. Each endopod seems to consist of 11 podomeres and a terminal element. On the 9th and 11th left endopods and the 9th and 10th right ones, which have been partly or completely exposed by preparation, the two proximal podomeres appear particularly robust, and each is longer than wide. Nine additional short podomeres are discerned on the 9th

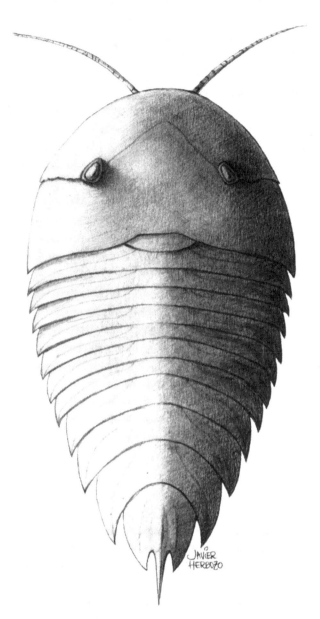

Fig. 73. Xandarella spectaculum Hou *et al.*, 1991. Reconstruction in dorsal view.

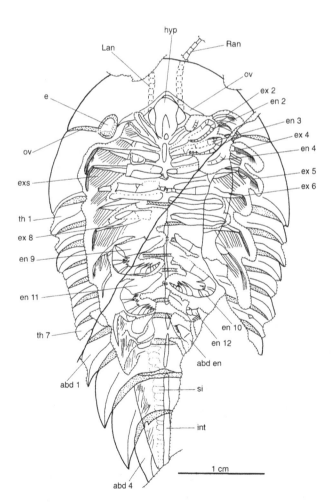

Fig. 72. Xandarella spectaculum Hou *et al.*, 1991. Drawing of CN 115286a, after preparation. For abbreviations, see Fig. 9.

left and 10th right endopods, both of which are completely exposed. Where the endopods are flexed, double transverse lines at the junction of two podomeres (Figs. 69, 72) possibly represent lines of overlap between two successive podomeres.

The three revealed thoracic endopods show the same ventral structure. Each of the podomeres from the 5th to the last one has a ventral spine, those on the 5th and 6th podomeres being long and strong and the following ones being gradually shorter distally. Clusters of 4–5 long ventral spines occur on the 3rd and 4th podomeres at the position of maximum bend of the appendage. Prepara-

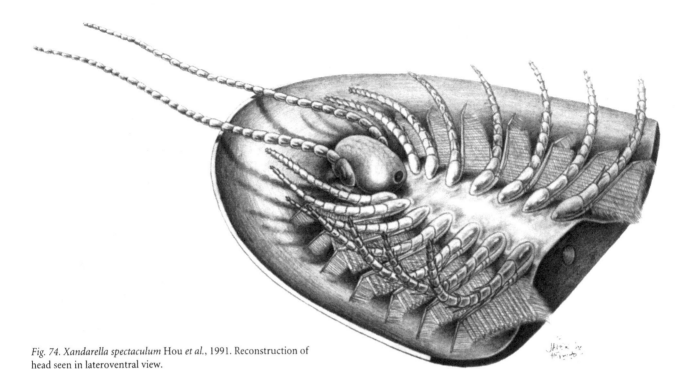

Fig. 74. Xandarella spectaculum Hou *et al.*, 1991. Reconstruction of head seen in lateroventral view.

tion did not reveal undoubted ventral spines on the two proximal robust podomeres (Fig. 72).

Preparation (Fig. 72) revealed dorsal spines on the 9th and 11th left and the 10th right thoracic endopods. (As mentioned above, they may have a basal articulation, if the variation in posture is not due to breakage). On the 9th left endopod, the long dorsal spines are indicated on the 2nd, 4th, 5th and 6th podomeres. The dorsal spine on the 2nd podomere appears to be longer than the podomere, and those on the 5th and 6th podomeres are more than twice the length of the corresponding podomere. The dorsal spines on the following podomeres tend to be gradually shorter. Most of the spines are preserved close to the dorsal margin of the endopods, but on the 12th right endopod, where the proximal portion is broken and only a little distal portion is revealed by preparation, a dorsal spine stretches from its base 45° out from the podomere. The terminal element of the 10th right endopod seems to bear two small spines flanking the terminal segment, suggesting the presence of dorsal and ventral spines on the terminal segment.

Neither exopods nor endopods of the abdomen are completely exposed by preparation. The distal portion of the 1st right abdominal endopod is broken. Its prepared proximal portion has many small ventral spines on the 1st podomere (Fig. 72).

Discussion. – The dorsal reconstruction of the animal (Fig. 73) is based mainly on the holotype, which is completely and symmetrically preserved (Figs. 70, 71A). The three-dimensional reconstruction is based on CN 115404

(Fig. 68A, B, 71F, G), which preserves appreciable convexity despite its thin exoskeleton.

The proximal part of the 6th and 7th right exopods in the holotype (Figs. 70, 71A) appears different from that of the 4th and 5th left exopods in CN 115286, where the proximal portion seems to be narrower and to widen abruptly in the distal end (Figs. 68C, 69A–B, 72). According to our interpretation, the main part of the two proximal segments have two wide posterior flanges, like those of the third segment. When the upper flange is destroyed, we see just the comparatively narrow axis. Distally, the axis of the second segment is extended posteriorly. and the flanges correspondingly disappear.

The structure of the 4th podomere of the exopods is somewhat less clear; it has an upper row of thin bristles and a lower row of flat setae, but it is unknown if there are additional structures under the setae. The tilt of the setae is less clearly seen than in the 3rd podomere.

Although the endopods of the head and thorax are well preserved and well exposed (by preparation, in CN 115286), some details are still uncertain. For instance, this is true regarding the number of endopod podomeres in head and thorax. On the proximal portion of the 6th left endopod of the head, there are two podomere boundaries shown as distinct furrows. However, there are also more obscure structures (wrinkles?), easily mistaken for podomere boundaries, which makes counting of the podomeres difficult. The same situation occurs on other endopods. Thus, for example, on the 2nd and 3rd right endopods of the head shield, there are more than 20 transverse lines, including podomere boundaries, ridges, folds and the like. In our counts, we have been careful to recognize only undoubted podomere boundaries.

The number of the endopods under the thoracic tergites is based on the 9th left endopod and 10th right one. An undoubted proximal segmental boundary divides into the two robust proximal podomeres that are longer than wide. Behind are the nine additional short podomeres and a terminal structure. Some weak transverse lines on the endopods may be interpreted as indication of additional podomeres; however, it is difficult to distinguish between occasional marks and real boundaries.

Although the endopods are fairly similar throughout the body, there is a distinct tagmosis. Similarities between head and thorax include the development of two or three notably long proximal podomeres, of which the first lacks ventral spines. Among notable differences, the thoracic podomeres are sturdier than those of the head and carry longer ventral spines and dorsal spines (or setae). In the thorax, the two podomeres at the maximum bend of the endopod carry a group of strong ventral spines, while there are no corresponding spines in the head. It is not known whether the limb corresponding to the small tergite behind the head shield is similar to the limbs in front or those behind. What is known about the abdominal appendages indicates that they differ from those of the head and thorax at least in having ventral spines on the 1st podomere.

In CN 115286, the three-dimensionally preserved gut appears to extend to the posterior end of the convex axial region (Figs. 68D, E, 71B, D), which is separated from the posterior margin by a pleural field (Figs. 68B, 70, 71A, F). The termination of the axis is where the anus was most probably situated.

Family Almeniidae n. fam.

Name. – After the nominal genus *Almenia*.

Diagnosis. – Xandarellids without compound eyes dorsally, with a large undivided head tergite, and with notably long overlap between tergites.

Genus included. – *Almenia* n. gen.

Genus *Almenia* n. gen.

Name. – After Mr. Sune Almén, one of our sponsors.

Type species. – *Almenia spinosa* n. gen. et sp.

Diagnosis. – As for the family.

Discussion. – The genus resembles *Xandarella* in the dorsal outline and large head shield, differing in the absence of dorsal compound eyes and sutures. In *Xandarella* there are four long abdominal tergites, which are not evident in *Almenia*.

Almenia spinosa n. gen. et sp.

Figs. 75–76

Name. – Latin *spinosus*, thorny, referring to the spinose tail.

Holotype. – CN 115405, a complete specimen with lower part and posterior portion of upper part, collected from level M3 in the quarry on the west slope of Maotianshan. No other specimens are known to us.

Description.. – (a) General characteristics: Although greatly different in detail, this species is notably similar in outline to *Xandarella spectaculum*: the head shield is large and the widest part of the animal; the body tergites taper steadily from the first thoracic tergite. The animal is about 30 mm wide (extrapolated from half the width) and 50 mm long. The tergum appears to have been evenly vaulted, and wrinkling indicates that the vault was high.

(b) Head: The anterior end of the head shield is not completely preserved, but the shape is semicircular. The head shield is estimated to be about 40% of the length of the entire animal and featureless; it lacks eyes, sutures, and topographic features, and its margin is smooth. The posterolateral corner is not preserved but appears to have extended into an acute genal angle (Figs. 75B, 76).

(c) Body: The body has ten tergites in addition to a tail shield (Fig. 76). One pair of spines in the tail shield indicates the presence of at least one segment in the tail, in addition to the telson. The tergites have considerable overlap between each other, are of subequal length, and have pointed lateral spines, particularly in the more posterior part of the body (Fig. 75A). The surface is smooth. It is not known if there is a small anterior trunk tergite, as in some specimens of *Xandarella spectaculum*.

(d) Ventral side: The posterior part of the hypostome is preserved. It seems similar to that of *Xandarella spectaculum* but probably lies closer to the anterior margin of the head. The antennal attachments are lost and the antennae unknown.

A number of legs have been prepared. The head has at least six pairs. In the endopods, the boundaries between the podomeres are fairly indistinct, particularly distally, but there appears to be some eight or nine podomeres. The distalmost ones are short. At least the distal podomeres carry a distal ring of fine bristles. In the head, the endopods appear to be cylindrical, whereas they are probably flattened in the posterior part of the body. The body appendages carry ventral spines. Toward the posterior end of the body, the appendages are very short and slender.

Fig. 75. Almenia spinosa n.gen. et sp. Holotype, CN 115405, from Maotianshan, level M3, ×3. □A. Upper part. □B. Lower part, prepared to show appendages. Note the large number of appendages in the head.

The exopods are also poorly preserved and commonly not even found between the dorsum and the endopods during preparation. However, it is clear that there is a comparatively long proximal podomere carrying lamellar setae and a shorter distal podomere, which is angled backwards (Fig. 76). Thus, the basic morphology is similar to that of *Xandarella spectaculum*.

A fairly narrow intestinal canal can be traced from between the first legs to posteriorly in the tail. It is partly filled with sediment. The termination is damaged, so that the position of the anus cannot be determined (Fig. 76).

Discussion. – Similarities with *Xandarella spectaculum* that imply a relationship include general body outline and

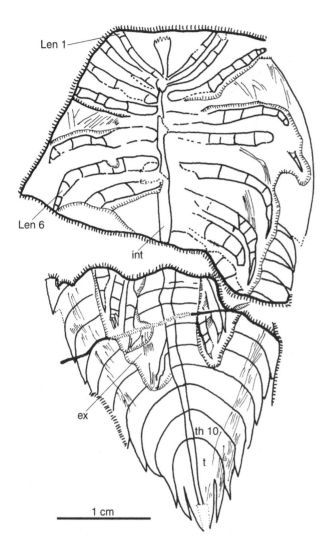

Fig. 76. *Almenia spinosa* n.gen. et sp. Drawing of lower part of CN115405, after preparation. For abbreviations, see Fig. 9.

convexity, a large head, the shape of the tail, the featureless surface, a large number of head appendages, the morphology of the endopods, and the two-segmented exopod. Features that are different in the new species include the reduced exopods, the lack of dorsal compound eyes and sutures, and the lack of very long posterior tergites. Actually, the tail tergite most probably includes at least one segment in addition to a telson. Also, the two tergites just in front of the tail tergite are a little longer than the preceding ones and may be composed of more than one segment, particularly if the small size of the posterior legs are taken into consideration.

Order Sinoburiida n.ord.

Diagnosis. – Petalopleuran arthropods of trilobite-like habitus, including separation into three tagmata; overlapping free thoracic tergites; trilobation due to the formation of an axial lobe; dorsal orientation of the visual surface of the compound eye. Mineralization and facial suture lacking.

Families. – Family Sinoburiidae n.fam.

Family Sinoburiidae n.fam.

Diagnosis. – As for the order.

Genus included. – *Sinoburius* Hou, Ramsköld & Bergström, 1991.

Genus *Sinoburius* Hou, Ramsköld & Bergström, 1991

Type species. – *Sinoburius lunaris* Hou, Ramsköld & Bergström, 1991

Diagnosis. – Head shield large, crescentic in outline, with paired eyes placed mediolaterally; antennae and four pairs of legs under head shield; thorax has seven tergites, each carrying one pair of biramous legs; subquadrate tail shield with terminal and two lateral pairs of spines, ten segments under tail shield, first six of which bear leg pair, succeeding four segments without legs.

Sinoburius lunaris Hou, Ramsköld & Bergström, 1991

Figs. 77–79

Synonymy. – ☐1991 *Sinoburius lunaris* sp. n. – Hou, Ramsköld & Bergström, p. 403, Fig. 4.

Holotype. – CN 115287 (Hou *et al.* 1991, Fig. 4), a complete specimen including upper and lower parts, from Maotianshan, level M2.

Other specimen. – One complete paratype specimen including upper and lower parts (CN 115288).

Distribution. – Known only from Maotianshan, level M2.

Description. – (a) General characteristics: The animal is less than 10 mm long. The original convexity of exoskeleton is partly preserved; it is more convex in the axial region. The animal is broad and large anteriorly, narrow and long posteriorly. The axial region is well defined by a shallow furrow, extends from the anterior head shield to posterior tail shield, is widest at the first thoracic tergite, gradually narrows toward both ends (Figs. 77, 78).

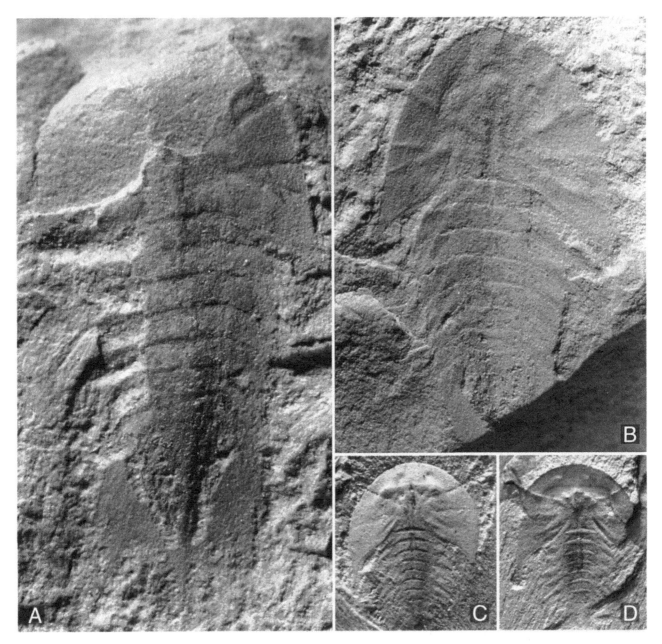

Fig. 77. Sinoburius lunaris Hou et al., 1991. □A, B. CN 115287, from Maotianshan, level M2. □A. Lower part after preparation, ×16.6. □B. Holotype, upper part, ×13. □C, D. CN 115288, from Maotianshan, level M2, lower and upper parts, respectively, ×8.3.

The holotype is 8.6 mm long and 4.4 mm wide; the paratype specimen measures 6.5 mm in length and 4.1 mm in width.

(b) Head: The head shield is large, crescentic in outline, about 1.5 times the width of the body. The posterior margin of the head shield is deeply embayed; long genal angles are present. Although compressed, the central part of the head shield clearly shows the development of a convex axis. A pair of large ovate compound eyes occurs medio-laterally.

(c) Body: The body is rectangular in dorsal aspect, consisting of seven thoracic tergites and one broad tail shield. The first tergite is narrowest, the second widest. Thereaf-

ter the body tapers only a little, all thoracic tergites behind the first one being of subequal width. There is a distinct imbrication of all tergites from head to tail (Figs. 77, 78). Each tergite has pointed pleural extremities which are long except in the first tergite. In each tergite, a ridge extends from the anterior end of the shallow axial furrow to the spine tip.

The broad tail shield resembles the tail shield of *Kuamaia lata* in bearing two pairs of lateral spines and one posteromedial spine. The anterior pair of lateral spines appears to be shorter than the posterior pair. The convex axial region ends close to the base of the terminal spine (Fig. 78A).

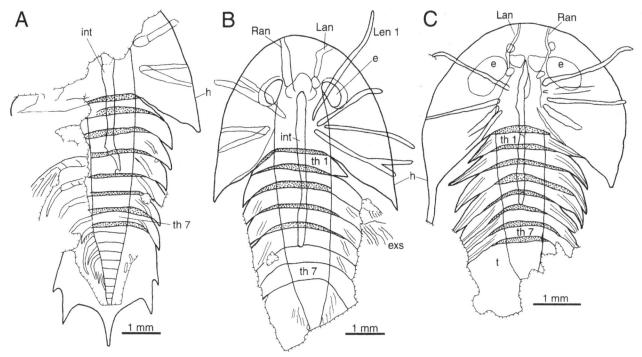

Fig. 78. *Sinoburius lunaris* Hou et al., 1991. □A, B. Drawings of lower and upper parts of CN 115287. □C. Drawing of lower part of CN 115288. Scale 1 mm. For abbreviations, see Fig. 9.

(d) Ventral side: In the paratype (Fig. 77C, D, 78C), one pair of anteriorly directed appendages, traced as shallow furrows on the head shield in the lower part, can been discerned and appears to represent antennae. In the holotype (Fig. 77B, 78B), the antennae are attached to the sides of a sub-triangular convex structure, which may represent the hypostome. The gut, partly filled with sediment, is seen from the tip of the hypostome in the holotype and can be traced to the base of the hypostome in the paratype.

In the upper part of the holotype, the antenna on the right side is obscure, although its site of attachment is visible (Figs. 77B, 78B). Four laterally directed biramous appendages, which extend far beyond the lateral margin of the head shield, seem to represent the legs behind the antenna. In the lower part of the paratype (Figs. 77C, 78C), four legs are also indicated by shallow furrows especially on the right side of the head shield. These four pairs appear to be the full set of head legs.

Preparation of the holotype has demonstrated one pair of biramous legs under each thoracic tergite (Figs. 77A, 78A). The exopod setae appear to be similar to those of *Naraoia longicaudata*. However, the detailed structure of the thoracic legs is not known, because the single specimen that can be prepared is not well enough preserved.

Segments in the tail are distinctly shorter than in the thorax (Figs. 77A, 78A). There is indication of about ten segments, the first six with paired legs, the rest without legs. The first left leg shows posteriorly directed setae, indicating that also the tail legs are biramous. These legs are bent strongly rearwards.

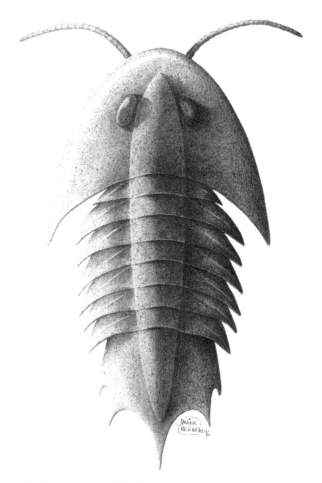

Fig. 79. Reconstruction of *Sinoburius lunaris* Hou et al., 1991, in dorsal view.

Discussion. – The reconstruction of the animal in dorsal aspect (Fig. 79) is based on the holotype. Even after compression, the holotype exhibits a notable convexity, particularly of the axis. The paratype is more flattened, so that the head shield appears to be broader than that of the holotype.

Sinoburius is reminiscent of trilobites because of apomorphic trilobation, the tripartite tagmosis, and the number of cephalic segments. It differs apomorphically in the dorsum in having merostome-type eyes looking dorsally. In lacking mineralization and facial sutures, it is plesiomorphic where trilobites have apomorphic modifications. The similarities may not indicate any close relationship within the Lamellipedia, as the derived character states were developed more than once within the group. Still, *Sinoburius* appears more similar to trilobites than do *Naraoia* and *Tegopelte*, which lack both trilobation and tripartite tagmosis.

Subclass Trilobita Walsh, 1771

The trilobites can be defined as follows: Lamellipedians with trilobation; pleural field extending around the body, which tends to be divided into the tagmata cephalon, thorax and segmented pygidium; tergum including eye lenses calcified; ventral side calcified only in doublure and hypostome; circumocular and submarginal sutures (probably succeeded by facial sutures) facilitating ecdysis; compound eyes dorsal, but not primarily looking dorsally through rounded opening in head shield as in other lamellipedians, but laterally through fissure; eye ridge primarily connecting eye with glabella, may be lost in later forms; behind the antennal segment there are four cephalic segments.

Despite having several distinctive features, trilobites have recently become confused with other lamellipedians, e.g., *Naraoia* (Whittington 1977) and *Tegopelte* (Whittington 1985a). In the case of the latter, Whittington (1985a, p. 1254) states that the 'filaments' of the exopod 'are like those of a trilobite ...', and *Tegopelte* is therefore placed in Class Trilobita'. Another cause for confusion is that the number of head segments was considered important (and thought to be identical). Actually, *Naraoia* may have one segment less (antennal plus three leg segments) than the original number in trilobites (with indication of four postantennal segments already in Lower Cambrian forms, and of 3–4 pairs of legs in younger forms, see below), and *Tegopelte* seems to lack a distinct head tagma (cf. Ramsköld *et al.* 1996). We find it more practical to accept the relationship between trilobites, *Naraoia*, *Tegopelte* and a whole suite of other lamellipedians (with 'filaments') on a higher taxonomic level and to restrict the term Trilobita to trilobites as traditionally understood. The obvious alternative would be to include all lamelli-

pedians in the Class Trilobita, but that approach would create more problems than it solves.

There is some confusion about the segmentation in the trilobite head. Only in *Rhenops* have we been able previously to recognize four pairs of legs behind the antennae (Bergström & Brassel 1984). In other trilobites, ranging from the Middle Cambrian to Devonian, only three pairs have been seen (Stürmer & Bergström 1973; Cisne 1975; Whittington 1975). The presence of three pairs in most trilobites might have been taken as evidence that this number is the correct one. However, the clear presence of four successive legs in a nicely preserved specimen is undeniable positive evidence, whereas the inability to find more than three pairs in other, less well-preserved specimens may be explained in different ways. It is possible that there were only three pairs in some trilobites, a state perhaps achieved by reduction of the frontal pair, which would seem to be almost inconveniently far in front. Alternatively, three may be the original number. However, in many trilobites with clear segmentation in the cephalic tergum, the antenna can be associated with the antennal pit and therefore with the fifth cephalic segment from behind (Henriksen 1926; Bergström 1973, pp. 9–12).

As seen below, in *Kuanyangia* it is clear not only that there are four glabellar segments behind the antennal segment, but also that there are four pairs of cephalic legs. Thus, whether or not there was any leg reduction in trilobites, antennal plus four cephalic segments appears to be the ordinary number of segments.

The supposed trilobite *Agnostus pisiformis* has very few characteristics that are trilobitan but shows features more typical of crustacean-like arthropods (Müller & Walossek 1987; Walossek & Müller 1990). Thus, agnostids do not seem to be at all related to trilobites. Eodiscids have always been considered intermediate between agnostids and 'other trilobites'. Now it is clear that eodiscids are trilobites, and therefore unrelated to agnostids.

Order Redlichiida Richter, 1933

Family Redlichiidae Poulsen, 1927

Genus *Kuanyangia* Hupé, 1953

Kuanyangia sp.
Figs. 80–81

There are a few trilobite species in the Chengjiang fauna. Shu *et al.* (1995) described and Ramsköld & Edgecombe (1996) discussed appendages in *Eoredlichia intermedia* (Lu, 1940). We have identified appendages in some trilobite specimens, of which two are of the redlichiid *Kuanyangia*. In the Lower Cambrian Qiongzhusi Formation,

Fig. 80. The redlichiid trilobite *Kuanyangia* sp., lower (A) and upper (B) parts of CN 115407, Maotianshan, level M2. ×1.5.

Eoredlichia Zone, specimens of *Kuanyangia* have been referred to as *Redlichia pustulosa* Lu, 1941, *Kuanyangia granulosa* Zhang, 1966, *K. wutingensis* Luo, 1975, *K. shishanensis* Luo, 1975, and *K. bella* Chen, 1983. Several or even all of these may belong to a single species. It is beyond the scope of this contribution to revise trilobite classification, and we therefore leave the determination as *Kuanyangia* sp. (Fig. 80). The following description excludes the mineralized exoskeleton.

Specimens. – CN 115407 from Maotianshan, level M2, one unnumbered specimen with appendages, and some unnumbered specimens that may or may not have remains of appendages.

Appendages. – One of the specimens was partly prepared from the dorsal side to show appendages in the cephalon and thorax (Fig. 81). Anteriorly is a pair of antennae. They have a fairly stout proximal portion. Beyond that follow more than 20 annuli (Fig. 80B). The distal part is missing, but the thickness at the broken end indicates that there are

many additional annuli. The antennae attach under the front lobe of the glabella, leaving four head segments behind (Fig. 81).

Behind the antennae, three cephalic appendages are clearly seen on the right side, plus a narrow fragment anteriorly which may represent a fourth limb. On the left side there are four cephalic exopods. In front of them there is a slim segmented appendage, which may either be a slim 1st endopod or an appendage belonging to another animal. It is distinctly narrower than the antenna. It is possible that this is the endopod of the 1st post-antennal appendage. In the successive cephalic and thoracic legs, the large basis has a finely serrated medial and ventral margin. Beyond the basis, the endopod consists of five or six additional, cylindrical podomeres. The terminal podomere carries a few strong but short spines. The exopod is a large blade consisting of two segments and tightly beset with setae along the entire posterior and distal margins. The setae are narrow but flat, having much the same appearance as in many other lamellipedians. As in *Naraoia*, the setae tilt towards the midline of the trilobite.

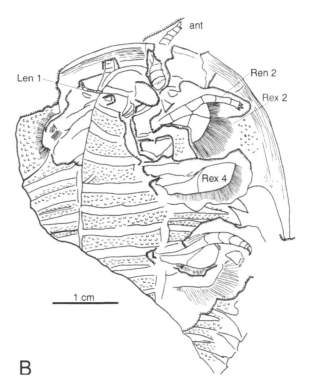

Fig. 81. Kunayangia sp. □A. Lower part of CN 115407, after preparation, × 1.6. □B. Drawing of the same. For abbreviations, see Fig. 9.

Other arthropods relevant to the discussion

This section discusses other fossil arthropods that are relevant to the Chengjiang fauna. Systematic placements of these arthropods are proposed under 'Summary of classification' below.

Crustaceans

Class Crustacea Pennant, 1777

Moore & McCormick (1969) rank the Crustacea as a superclass, and its subgroups are generally regarded as classes. Others have raised ranks even more. Thus, for instance, Manton (1978) assigned phylum rank to the Crustacea, and Abele *et al.* (1992, p. 374, Table 1) regarded crustacean subgroups (e.g., Maxillopoda) as superclasses. However, considering the related non-crustacean forms in the Palaeozoic, we need systematic space above the level of the Crustacea. The Crustacea must be considered as only one subgroup of the Crustaceomorpha, which in turn is a subgroup of the Schizoramia, which may be a subgroup of arthropods. On the whole, the systematic ranks are high within the Crustacea, if compared with other arthropods, and too high to accommodate new finds of Cambrian crustaceans without making a new class for each of them.

In short, and based on both morphological and RNA sequence data, we feel that the Crustacea are better treated as a class, which may be subdivided into five subclasses, *viz.* Cephalocarida, Branchiopoda, Maxillopoda, Branchiura, and Malacostraca. In the Cambrian, the branchiopods appear to be represented by, among others, *Branchiocaris*, *Odaraia* and *Rehbachiella*, and the maxillopodans by *Bredocaris*, *Skara*, *Dala* and *Walossekia*. Remains of additional probable branchiopod forms are known from the Lower Cambrian of Canada (Butterfield 1994). Branchiurans (pentastomids) appear to be present in the Lower Ordovician (Andres 1989). All of these forms, except for *Branchiocaris* and *Odaraia*, are from the Alum Shale sequence in Sweden. Cephalocarids, branchiurans and malacostracans have not been identified with certainty in the Cambrian. In addition to these forms, there are non-branchiopod meta-nauplius larvae in the Alum Shale (Müller 1981; Fryer 1985, p. 111).

Schram's (1986, pp. 542–544) radical treatment of crustacean systematics created confusion. For instance, his Subclass Phyllopoda (=Branchiopoda) includes both cephalocarids and phyllocarids, the latter embracing both non-crustaceans (Canadaspidida) and malacostracan crustaceans (Archaeostraca, Leptostraca). The malacostracan nature of archaeostracans and leptostracans is beyond doubt (Bergström *et al.* 1987; Dahl 1987).

Branchiopods

Subclass Branchiopoda Latreille, 1817

It is not generally accepted that there are branchiopods in the Lower or Middle Cambrian, although *Canadaspis* has been accepted as a member of the more derived malacostracan crustaceans. The presence of branchiopods in the Upper Cambrian is not in doubt after the description of *Rehbachiella* (Müller 1983; Walossek & Müller 1992; Walossek 1993).

In an attempt to recognize additional Cambrian branchiopods, it is the characteristics of the body or carapace that are of practical importance; the legs are rarely preserved or easy to interpret. One such character, found among the Notostraca, Kazacharthra and some Conchostraca, is a large telson in combination with extremely short trunk segments (or rings). Cambrian arthropods with this combination include *Protocaris*, *Branchiocaris*, *Odaraia* (Fig. 38), and perhaps *Banffia* (if the latter is an arthropod). To this list we want to add the recently described genus *Vladicaris* (Chlupáč 1995) from the Lower Cambrian of the Czech Republic. Briggs (1976 and 1981) compared *Branchiocaris* and *Odaraia* with branchiopods but did not find evidence for their inclusion even within the Crustacea. Simonetta & Delle Cave (1975, pp. 31–32) had previously placed the two genera (*Branchiocaris* as 'Protocarida') close to conchostracan and perhaps cladoceran branchiopods. Whittington (1979, p. 258) could not place *Branchiocaris* in any group of living arthropods and did not judge the systematic position of *Odaraia*. Briggs & Fortey (1989, Fig. 1) placed *Branchiocaris* between the lamellipedian *Marrella* and the derived crustacean *Speleonectes* (belonging to the secondarily elongated maxillopod group Remipedia), while *Odaraia* was placed in a mixed group with *Waptia* (Pseudocrustacea), *Perspicaris* (Pseudocrustacea) and *Canadaspis* (Paracrustacea). In the same diagram, lamellipedians (except for *Marrella*) and chelicerates are derived from crustaceans. As can be seen, the literature gives a confused picture of relationships.

The hardness of the Burgess Shale makes preparation of the specimens extremely difficult. As far as we can tell, there is not a single arthropod leg that has been prepared to show its entire outline, with consequences for the interpretation. In *Branchiocaris pretiosa*, the legs are reconstructed as simple triangular flaps, which give no hint regarding relationships (Briggs 1976, Text-fig. 2). Yet the best specimen illustrated by Briggs (1976, Pl. 3) (cf. Fig. 82 herein) shows nicely preserved medial edges of the posterior legs (USNM 189028). Its endite lobes are very similar to the legs in many branchiopods (for comparison, see Schram 1986). Similar appendages are not known from any non-crustacean arthropods. Both 1st and 2nd antennae are uniramous, which is rarely the case in crustaceans, except for adult notostracans and anost-

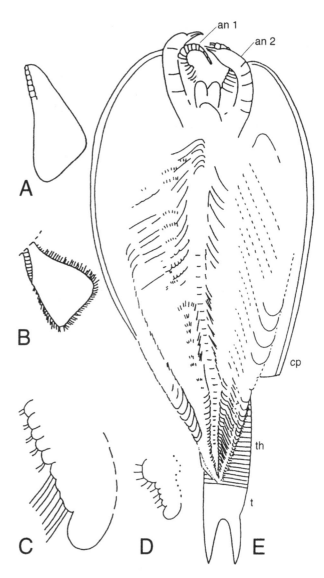

Fig. 82. Branchiocaris pretiosa (Resser, 1929) from Burgess Shale, probably a calmanostracan branchiopod. □A. Body appendage as interpreted by Briggs (1976, Text-fig. 2). □B. Body appendage as interpreted by Delle Cave & Simonetta (1991). □C, D. Our interpretation: C, anterior body appendage; D, posterior body appendage, in part based on USNM 189030, in which appendages are turned forwards. □E. Entire animal, USNM 189028. For abbreviations, see Fig. 9.

racans. Probable eye lobes (Briggs 1976, Pl. 3c, d) occupy the same position as the eyes in *Rehbachiella* and several other primitive crustaceans. In addition, the entire habitus is that of a branchiopod, and we do not doubt that *Branchiocaris* is a true branchiopod related to notostracans and kazacharthrans.

Odaraia alata was similarly difficult to prepare and interpret. Nevertheless, the reconstruction presents more detail than available for *Branchiocaris pretiosa*, in particular for the distal part of the endopod, which is branched (Briggs 1981, Fig. 103). There is seemingly conflicting evidence for the proximal part of the endopod. In Briggs' Figs. 22, 23 and 26, the supposed endopods appear as long strings of beads. This is a reasonable appearance of the medial edge of a branchiopod appendage (see, e.g., Fryer 1985, Fig. 3). In Briggs' Figs. 77, 78 and 87, on the other hand, the endopod is a broad, flat flap. Along the posterior (i.e. inner) margin there is, however, an endite row that looks like a narrow segmented limb if seen in isolation. In Briggs' reconstruction, the proximal part of the endopod is drawn as a thick, segmented ramus of equal width, quite unlike the flat flap with convex inner (posterior as seen) margin seen in his Figs. 77, 78 and 87. Given a large flat exopod and a flat branched endopod, the appendage is clearly similar to that of modern notostracans and anostracans and extinct kazacharthrans. No even remotely similar legs are known from any non-crustacean arthropods. Furthermore, *Odaraia* has a clearly defined head with slender appendages (Briggs 1981, Figs. 76, 86) as well as powerful mandibles (Briggs 1981, Figs. 1, 5, 7, 15, 18, 26, 29, 35). In Briggs' explanatory drawings, only the cutting edge of the mandible is interpreted as a mandible, while what appears as the main body of the mandible is interpreted as 'muscle scar on carapace'. This area may be sharply delimited morphologically (e.g., Briggs 1981, Fig. 6), and not only by a diffuse colour difference. As a whole, the mandible is closely comparable to that of notostracans.

We are therefore fairly confident that both *Odaraia* and *Branchiocaris* are true branchiopod crustaceans. The same must hold for *Protocaris* and *Vladicaris*, which are fairly similar to *Branchiocaris*. When we turn to genera known only from carapaces, we are on much more shaky ground, and the higher classification must be tentative (cf. Briggs 1978b, p. 482; Robison & Richards 1981).

The Upper Cambrian *Rehbachiella* (Müller 1983; Walossek & Müller 1992; Walossek 1993) is the most generalised branchiopod we know, and all others could be derived from such a source. In the Calmanostraca lineage, the next observable step is a strong crowding of the posterior trunk segments. In one branch, the Protocaridida, the 2nd antenna lost its endopod and may have been prehensile, as in male anostracans. In the Odaraiida, the carapace appears to have embraced also the ventral side, and at least in *Odaraia* the telson was produced into a fin (Fig. 38). The third branch includes the Notostraca, Kazacharthra and Diplostraca/Onychura (Conchostra and Cladocera). A characteristic feature is the development of the maxillary (or shell) gland within the carapace. A second diagnostic feature is the appearance of compound eyes internally on the dorsal side. In the Conchostraca the carapace is no longer shed during moulting. The Cladocera arose from conchostracans through miniaturisation.

Marrellomorphs

Class Marrellomorpha (Beurlen, 1934) Størmer, 1944

An emended diagnosis could read as follows: Lamellipedians with exopod podomeres, each carrying one large seta; without pleural folds on individual segmental sclerites and without mineralization of the tergum; appendages with semi-pendent stance.

Orders and genera included are the Marrellida Raymond, 1935, with *Marrella* Walcott, 1912; Mimetasterida Beurlen, 1934, with *Mimetaster* Gürich, 1932.

Størmer (1944, p. 134) attributed the name of the class to Beurlen, but this author erected the Marrellomorpha as an order (Beurlen 1934). We therefore regard Størmer as the author of the class name.

This group was discussed by Stürmer & Bergström (1976) and Bergström (1978, 1981). As we see it, the core of the class is formed by *Marrella* from the Middle Cambrian and *Mimetaster* from the Devonian (Stürmer & Bergström 1976). They are characterized by a large carapace-like head shield with pleura extending horizontally, lack of pleura in the body, and by semipendent legs with flagellar exopods, in which each segment carries one long seta (Bergström 1992).

One of us (e.g., Bergström 1978) has previously advocated that the Middle Cambrian *Burgessia* belongs within the Marrellomorpha. Hughes (1975) reconstructed the endopods of *Burgessia* as strongly laterally deflected and curved (Fig. 83A). The laterally deflected posture is a characteristic of many lamellipedians. However, most specimens seen in lateral view actually expose straight

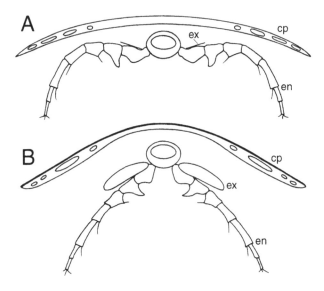

Fig. 83. Burgessia bella Walcott, 1912 from the Burgess Shale. □A. Section through trunk according to Hughes (1975, Fig. 10B), with limbs deflexed laterally as in trilobites. □B. Section through trunk, in our interpretation, with semipendent limbs as seen in many of the preserved specimens. For abbreviations, see Fig. 9.

endopods, which hang down from the body. In rare cases (e.g., Hughes 1975, Pl. 1:2, 3, 8, 11, 12; Simonetta & Delle Cave 1975, Pl. 58:9), a semipendent posture is clearly revealed (Fig. 83B). Widening of the angle between the two legs of a pair (as in Hughes' Pls. 4:2, 4; 5:2, 5) can be explained by the embedding angle, but the same cannot be true of narrowing of the angle. Simonetta & Delle Cave (1975, Pl. 8:1d) reconstructed the endopod with a reasonable stance, but with a marked knee not seen in the actual specimens. This means that *Burgessia*, which lacks lamellar setae, lacks characters tying it to the Lamellipedia and its systematic affiliation is open.

Acercostracans

Order Acercostraca Lehmann, 1955, of the Subclass Nectopleura, Class Artiopoda

Vachonisia (Figs. 84–85) is known only from the Devonian of Germany (Stürmer & Bergström 1976). Previously the Acercostraca Lehmann, 1955, with *Vachonisia* Lehmann, 1956, and questionably also the Superfamily Cycloidea Glaessner, 1928 (with a number of families, such as Cyclidae Packard, 1885 and Hemitrochiscidae Trauth, 1918), were referred to the Marrellida (cf. above). We now see distinct similarities between *Vachonisia* and nectopleurans, particularly perhaps with larval *Naraoia*, which presumably indicate affinities. The fusion of the abdominal tergum into one shield in *Naraoia* appears to be the result of paedomorphosis (Fortey & Theron 1994), and the fusion of the entire tergum in *Vachonisia* appears to be a further step on the same road.

Without having seen specimens or radiograph films, Whittington (1979, p. 259) claimed that there is no evidence for two leg branches in *Vachonisia*. This is not true: stereo radiographs (Stürmer & Bergström 1976, Pl. 18b) quite distinctly show the strongly bent endopods on both sides of the midline; even the segmentation is clearly visible. The exopods are best seen far back on the left side. Their shafts were seen and accurately drawn already by Lehmann (cf. Tasch 1969, Fig. 47). The exopod setae are distinctly seen in the original radiographs. Because the radiograph films are very soft, a great amount of information was regrettably lost in the process of making prints.

The type specimen of *Vachonisia rogeri* could not be found at the time of our description (Stürmer & Bergström 1976). It subsequently did re-appear and plainly exposes the features we knew from the radiographs (Fig. 84).

The Carboniferous to Triassic cycloids expose certain similarities to *Vachonisia*, although they also differ from this genus in various respects. Thus, they have an enlarged shield and semipendent appendages with flagelliform exopods. It appears possible that they are highly derived relatives of *Vachonisia*. Alternatively, they may be proschizoramians.

Tegopeltids

Family Tegopeltidae Simonetta & Delle Cave, 1975 of the Order Helmetiida, Subclass Conciliterga, Class Artiopoda

Diagnosis. – Lamellipedians with tergum completely fused into single shield. (No diagnosis has been presented previously.)

Simonetta & Delle Cave (1975, pp. 33–34) included the Tegopeltidae (with the genus *Tegopelte* Simonetta & Delle Cave, 1975) in their Order Tontoiida. Whittington (1985a, p. 1256) pointed out that, by mistake, Simonetta & Delle Cave (1975, Pl. 37:3) had illustrated the holotype of *Mollisonia symmetrica* as a specimen of *Tontoia kwaguntensis*. He argued that the latter is not a recognizeable taxon and is perhaps not even an arthropod. Delle Cave & Simonetta (1991, p. 199 and tab. I) rejoined that the species is valid, and that *Tontoia* and *Tegopelte* are closely related. Not having seen the type specimen in question, we cannot comment on the different views. *Tegopelte* superficially looks very like *Saperion*, but we cannot see any evidence for either rostral and pararostral plates or for a rostral doublure; we include it only tentatively in the Conciliterga.

Whittington (1985a, pp. 1254, 1273) thought that *Tegopelte* had a number of tergites and that these were merely separated by 'a faint line of junction' (if so, they would be no overlap between them). He further considered that the articulation between the supposed tergites is just as in the trilobite *Schmalenseeia* Moberg, 1903. However, *Schmalenseeia* had free segmental tergites, which overlap and apparently could slide over each other; *Tegopelte* differs in having no overlap between adjacent tergites, if indeed there are any free tergites, which we do not believe. Ramsköld *et al.* (1996) demonstrated that transverse lines such as the supposed tergite boundaries are artefacts and can be seen in large valves such as those of *Tegopelte* and *Naraoia*. The similarity therefore is very remote. Whittington (1985a, p. 1273) further argues that the absence of mineralization in *Tegopelte* would have made the tergum flexible; however, with a vaulted tergum the flexibility must have been virtually nil.

It is worth noting that Whittington (1985a, p. 1263) described the exopod setae of *Tegopelte* as probably oval in cross section, and oriented vertically to the surface of the exopod shaft. Thus shape and arrangement represent a clear similarity to the condition found in other lamellipedians.

Xenopodans

Subclass Xenopoda Raymond, 1935

The Subclass Xenopoda is considered to be a synonym of the Subclass Emeraldellida Størmer, 1944, pars, and the Subclass Prochelicerata Størmer, 1944, pars.

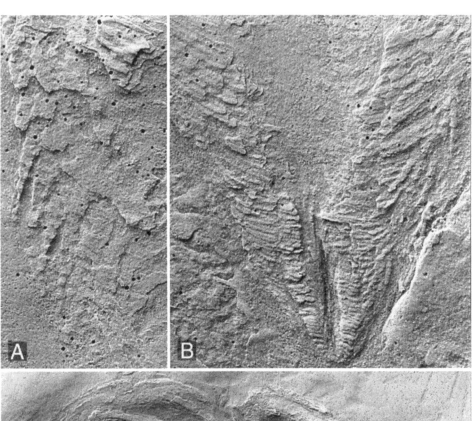

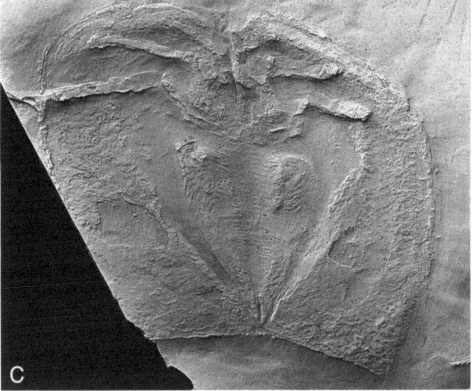

Fig. 84. Vachonisia rogeri (Lehmann, 1955), a marrellomorph lamellipedian from the Devonian Hunsrück Slate in Germany. Specimen described by Lehmann and curated in the Senckenberg Museum (radiograph number WS 4743). □A. Tips of multiarticulate exopod shafts (cf. left side of C). □B. Lamellar exopod setae, rear part of body. □C. Whole individual in ventral view; specimen about 65 mm long.

A diagnosis may read as follows: Artiopodans with overlapping segmental tergites, without mineralization in the exoskeleton, with or without legs in the head tagma, a large ventral rostral plate (or hypostome), abdomen without legs and pleural folds, telson flanked by uropods. Feeding not through mud ingestion.

Raymond included the single Order Limulava in this subclass, with the genera *Sidneyia*, *Amiella* and *Emeraldella*. *Amiella ornata* is now known to be an anomalocaridid (see Hou *et al.* 1995, Fig. 17). We tentatively accept the association of *Sidneyia* with *Emeraldella*, even though there are differences between them, such as in the

Fig. 85. Vachonisia rogeri (Lehmann, 1955). Radiograph (WS 2804), middle part of animal. Endopods seen on both sides of dark midline; they are bent so that the tips point to the midline. On the right side are seen multiarticulate exopods with setae hanging backwards–inwards. Height of figure about 7.6 mm.

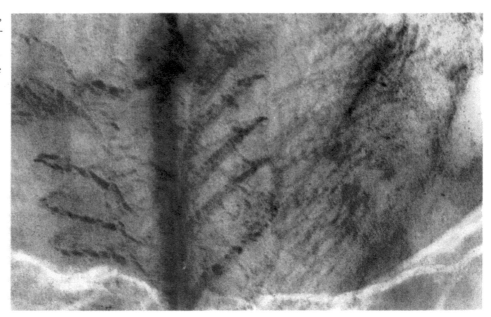

composition of the head tagma and in the leg morphology. The similarities between them are also shared by the aglaspidids, except for the offset abdomen. The postventral plates in aglaspidids, the paired flaps at the base of the tail spine in *Emeraldella*, and the appendages flanking the telson fin in *Sidneyia* may represent a 'character' uniting these arthropods. A similar arrangement is seen in *Cheloniellon* which, therefore, is tentatively included in the subclass. As mentioned previously, it was regarded by Størmer (1944) as belonging to a discrete subclass, the Cheloniellida.

There are certain similarities between xenopodans and aglaspidids. Members of both groups tend to have a smoothly vaulted tergum with 11–13 overlapping body tergites, and a semicircular head shield without sutures and without genal spines. *Sidneyia* appears to differ from the others in being more advanced in its appendages, but their structure is poorly understood, and as probable autapomorphies they do not tie *Sidneyia* to any other arthropods, nor separate it. Some of the similarities, such as the regular overlap between adjoining tergites, are plesiomorphies and therefore of restricted significance.

Emeraldellids

Order Emeraldellida Størmer, 1944

A diagnosis may read as follows: Merostomoids with fairly large, semicircular head shield, thorax with eleven free, overlapping tergites with pleura, and two ring-shaped sclerites which have a terminal spine. No eyes are known.

The group includes only the Family Emeraldellidae Raymond, 1935 with the genus *Emeraldella* Walcott,

1912. The only known species is *Emeraldella brocki* Walcott, 1912 from the Burgess Shale (Fig. 86).

The history of the taxonomic position of the Emeraldellidae is confused. Størmer's (1944, 1959) diagnoses of his Subclass and Order Emeraldellida state the possession of 'practically unaltered trilobitan appendages'. He also thought that the body was trilobed, which is not the case (Bruton & Whittington 1983).

The Order Emeraldellida and the Subclass Emeraldellida were both suggested by Størmer in 1944 (p. 134). Yet in 1959 (p. O30) he again erected the Order Emeraldellida, this time through translation from the Subclass Emeraldellida Størmer, 1944. Different animals have been included in the order. Størmer (1944) first included *Emeraldella*, *Molaria*, *Habelia* and *Naraoia*, but later (Størmer 1959) only *Emeraldella*. Delle Cave & Simonetta (1991) believed that they had authored the order in an earlier contribution (Simonetta & Delle Cave 1975), although in that paper they referred the Order 'Emeraldella' to Størmer. In 1991 they included *Emeraldella*, *Ecnomocaris*, *Habelia*, *Molaria*, *Sarotrocercus*, and *Thelxiope*, genera united only by the presence of a tail spine, in the Emeraldellida. *Emeraldella* is not similar to any of the other genera. Bruton & Whittington (1983) criticized these attempts but offered no solution to the problem.

In phylogenetic analyses, the closest relatives of *Emeraldella* have been thought to be either the megacheiran arthropods *Alalcomenaeus*, *Actaeus* and *Yohoia*, or the crustaceomorphs *Odaraia*, *Waptia*, *Canadaspis* and *Perspicaris* (Briggs 1983), or the lamellipedian *Sidneyia* and the proschizoramian *Sanctacaris* (Briggs 1990), or just *Sidneyia* (Briggs *et al.* 1992). Simonetta & Delle Cave (1975) regarded emeraldellids to be systematically close to

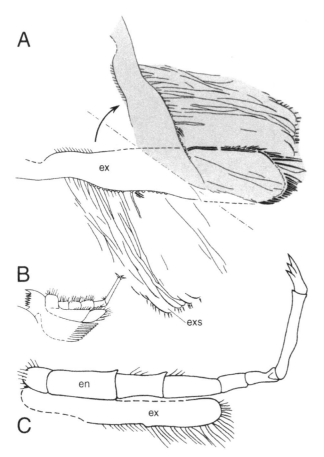

Fig. 86. Emeraldella brocki from the Middle Cambrian Burgess Shale. □A. Detached exopod illustrated by Bruton & Whittington (1983, Figs. 47, 49) and used for their reconstruction of the appendage (reproduced here in B). The shaded area shows the specimen as preserved. The proximal portion is folded (arrow) over the distal part. In the unfolded condition, the exopod is seen to be long and narrow and beset with posterior setae like the exopods in situ. □B. Bruton & Whittington's (1983, Figs. 64–65) reconstruction. In this, the proximal and distal parts of the exopod shown here in Fig. 86A are interpreted as two discrete branches. □C. Endopod and exopod as seen in fossil remains, but composed of parts from nearby limbs. The composition is based on Fig. 35 of Bruton & Whittington (1983). The endopod is put together from endopods nos 9 and 10 on the left side, while the exopod is no. 8 on the right side, with relative length decided from the comparison with the left side of the same segment. The hinge between the two branches is not seen, but is based on the authors' notion that the two branches are firmly held together and on a comparison with *Naraoia*. This kind of hinge forced the branches to move in unison, but enabled the endopod to swing forwards and backwards to the posture seen in the specimens. For abbreviations, see Fig. 9.

nectaspidids (*Naraoia*) and mollisoniids. Simonetta & Delle Cave (1975, p. 31) and Delle Cave & Simonetta (1991, Tabs. 2, 4), thought that emeraldellids gave rise to chelicerates and possibly to insects and myriapods. Bergström (e.g., 1992) held *Emeraldella* to be a trilobitomorph (i.e. lamellipedian).

In hindsight, most of the speculation on systematic position has been based on very poor or even faulty observations. Størmer's (1944, Fig. 17:3) drawing is reasonably good but not revealing much detail. Simonetta & Delle

Cave (1991) introduced a peculiar dorsal element at the base of the exopod. In the reconstruction of Bruton & Whittington (1983, Figs. 64–65), the exopod has become a unique bilobed, vertical element, unlike the exopod of all other arthropods both in morphology and posture.

Bruton & Whittington's reconstruction is based on a misunderstanding of the material. There are two detached exopods and a detached endopod thought to be attached to one of them (Bruton & Whittington 1983, Figs. 43–45, 47, 49). Both exopods consist of a large flat element with a notable bend (Fig. 86A). Such a bend is not present in the complete specimens, and one may wonder why the endopods have been assigned to *Emeraldella* at all. For instance, Bruton & Whittington's Figs. 30, 34, 36 and 38 clearly show how the 'filaments' are attached along a straight line, not on a boomerang-shaped structure as in one of the detached specimens. The other detached specimen does not have any 'filaments' at all in the corresponding position, but this was overlooked in the reconstruction. Furthermore, the long 'filaments' are along the posterior margin in the specimens (their Figs. 29, 30, 33, 34, 36–38) but distally in the reconstruction (Fig. 86B). The reconstruction also is clearly at odds with the straight exopods seen in the entire animals.

A possible explanation is that the two detached exopods are folded over on themselves (Fig. 86A, C). In this way, the posterior setae of the proximal part of the shaft are seen on top of the distal part of the shaft in Bruton & Whittington's Figs. 45 and 47 and thought to be a distinct branch (given as 'gb' in the reconstruction). As a result, the exopod was thought to be attached posteriorly to the endopod, not anteriorly as it should be. Still, it is difficult to understand why the entire limb is reconstructed as in a single plane. Undoubtedly, there was a flexure between endopod and exopod, such as is borne out, for example, by their Figs. 34 and 36, in which both endopods and exopods are folded backwards. Consequently, *Emeraldella* clearly had a limb of trilobite design, as maintained by Størmer (1944, 1959). Also, the difference, if any, between head and body legs was much less pronounced than stated by Bruton & Whittington.

Another observation from the material is that the exopod 'filaments' clearly are attached to the axis via an articulation (Bruton & Whittington 1983, Figs. 45) and, therefore, by definition are setae. The long posterior setae have marginal bristles. The short distal setae look like knife blades, with one thickened side and one thin edge. The distance between two setae is much smaller than the width of each. Hence it is clear that the surface of each seta was not in the plane of the flat exopod axis, but at an angle, as in trilobites and other lamellipedians. Even if the morphology of the long setae is less obvious, it appears that the flatness and arrangement is as in other lamellipedians, but perhaps less advanced – the setae appear thinner. They are preserved overlapping each other in the bedding plane. It is difficult to tell if they were twisted as much as 90° to the plane of the exopod shaft.

Ultimately, the Bruton & Whittington (1983, Figs. 61–64) reconstructions show an animal lifted horizontally on legs of sub-equal length, whereas specimens (for instance, their Figs. 33–34) show that the leg length decreased strongly backwards. The endopod in leg segment 8 appears to have been about three times as long as that of segment 16.

Limulavids
Order Limulavida Walcott, 1911

This group is based solely on the Family Sidneyiidae Walcott, 1911, with *Sidneyia* Walcott, 1911.

We have not recognized any limulavid in the Chengjiang fauna. The genus so far includes only *Sidneyia inexpectans* Walcott, 1911. Bruton (1981) concluded that its appendage structure is similar to that in modern xiphosurans and that *Sidneyia* may be close to the ancestry of chelicerates.

Sidneyia is difficult to interpret, in part at least because of preservational factors. General features indicate that it belongs to the lamellipedians. Thus, the endopods extend laterally, the outer branch has lamellar setae, the doublure widens anteriorly to a broad hypostome-like structure, and the eyes do not appear in front of the head shield. Because of the very short head tergite, the eye appears behind the head tergite, but it would be enclosed if more tergites were fused to the head. The next step in such a fusion appears to be present in *Xandarella*, in which the eye seems to pop up in a segmentary boundary which is not yet entirely fused.

The little evidence there is from the appendages is confusing (cf. Simonetta 1992). Bruton (1981, p. 650) thought that an extension from the basal podomere (supposed to be a coxa) was attached to the body side, much as in modern xiphosurids, while the 'gill' was attached to the second segment. He concluded that the trilobite 'coxa' is not homologous with the 'coxa' of *Sidneyia*. He did not discuss possible alternative interpretations, e.g., that the outer branches are not homologous, that the outer branch has shifted position, that the outer branch in fact attaches to the 'coxa', that the 'coxa' has split into two segments in *Sidneyia*, or that the extension from the 'coxa' is an exopod rather than an expanded part of the segment itself. The exact position of the attachment is difficult to see in *Sidneyia* (cf. Simonetta 1992), and to some degree it can shift, as shown by the extended attachment in *Naraoia*.

Contrary to Bruton's conclusions, there is no evidence that the exopod setae hang down between the endopods. His Figs. 84, 87, 90 and 91 show whole series of exopod setae ('gills') without endopods in between, and he actually stated that 'when found in situ on specimens preserved in parallel aspect, filaments of the gills always lie directly beneath the dorsal exoskeleton' (Bruton 1981, p. 653).

Bruton (1981) documented in some 20 illustrations a large flap, which he called a gill flap and which is probably part of an exopod. The attachment of the setae, Bruton's gill filaments, is not seen in any specimen. Bruton suggested that they are attached to the anterior margin of the flap. However, this margin is never seen, nor is there any flap associated with isolated bundles of setae. In certain cases, such as in Bruton's Fig. 91, the setae appear to diverge from an axis extending posterolaterally rather than anterolaterally, as the anterior margin of the plate appears to do. These conditions indicate that the setae were not attached to the plate at all, but to a narrow shaft, approximately as suggested by Simonetta & Delle Cave (1975, Pl. 7:1e). It is possible that the big flap and the shaft with setae, if there was such a shaft, were both parts of the exopod. The detailed arrangement remains unknown.

Even if *Sidneyia* is insufficiently known and understood, it is clear that it is very different from, for example, trilobites. The morphology of the appendages appears strongly derived, whereas, for instance, the shortness of the head tagma is less derived than the condition in trilobites. None of the conditions reveals anything about relationships. Even if Bruton's interpretation of the legs turns out to be correct, it does not support his conclusion for a chelicerate affinity, since the supposed similarity is between *Sidneyia* and the strikingly derived xiphosurids, not between *Sidneyia* and primitive merostomes. Thus, if there is a similarity, it is possibly caused by parallel adaptation to crushing of hard food and not to a shared origin of morphology. The exopod plate is a special autapomorphic adaptation in *Sidneyia* and has no bearing on its systematic affinity.

Without an understanding of the construction of the exopod, but with evidence of flat setae of lamellipedian type, there is no reason to claim that *Sidneyia* is fundamentally different from trilobite-like arthropods (cf. Bruton 1981, p. 649).

Cheloniellids
Order Cheloniellida Broili, 1933

Included in this order is the Cheloniellidae Broili, 1932, with *Cheloniellon* Broili, 1932. The genus so far includes only the Devonian *Cheloniellon calmani* Broili, 1932 (redescribed by Stürmer & Bergström 1978). *Cheloniellon* is clearly a lamellipedian, having dorsal, sessile eyes, laterally deflected legs, lamellar setae, and a doublure. It is more advanced than any lamellipedians in having two pairs of preoral appendages, namely one pair of flagelliform antennae and one pair of uniramous, probably grasping appendages. Perhaps connected with this development, there is no real hypostome: the spiny plate in front of the mouth is not elevated. With the loss of the antennae, and with the addition of one or two segments to the head tagma (Stürmer & Bergström 1978, Table 2), *Cheloniellon* would qualify as a chelicerate. From a morphological point of view, *Cheloniellon* is in some respects an almost perfect intermediate between lamellipedians and chelicerates. However, it is of course much too young to be likely as a phylogenetic intermediate, and it has lost the primitive tail sclerite, which is preserved in aquatic chelicerates. *Cheloniellon* is the best example to show how easily lamellipedians may have given rise to chelicerates.

Broili (1932) regarded *Cheloniellon* as a crustacean, Boudreaux (1979, p. 119) placed it next to the crustaceans, and Størmer (1959, p. O35) and Stürmer & Bergström (1978) placed it among the trilobitomorphs (i.e. lamellipedians).

Aglaspidids

Subclass Aglaspidida (Walcott, 1911) Bergström, 1968 (nom. correct. herein, ex Subclass Aglaspida Bergström, 1968)

Emended diagnosis. – Artiopodans with mineralized exoskeleton, with eleven overlapping segmental tergites behind the head, and a tail spine. Uropods present. Feeding not through mud ingestion.

Included are the Order Aglaspidida Walcott, 1911 (correct name first used by Briggs *et al.* 1979, without formal correction, ex Order Aglaspina Walcott, 1911; Hesselbo 1992 thought that Raasch 1939 was the author of the ordinal), containing the families Aglaspididae Miller, 1877 and Beckwithiidae Raasch, 1939 (but see below), and Order Strabopida n.ord., containing the families Strabopidae Gerhardt, 1932 and Lemoneitidae Flower, 1968.

No aglaspidid is known from the Chengjiang fauna. The American Aglaspididae were recently redescribed by Hesselbo (1992). Eurasian genera that have been referred to the Aglaspididae include forms from the Ordovician of Asia and Bohemia that have been studied by Chernyshev (1945, 1953), Andreeva (1957), and Chlupáč (1963a, b), namely *Angarocaris* Chernyshev 1953, *Chacharejocaris* Chernyshev, 1945, *Girardevia* Andreeva, 1957, *Intejocaris* Chernyshev, 1953, *Obrutschewia* Chernyshev, 1945, *Schamanocaris* Chernyshev 1945 and *Zonozoe* Barrande, 1872. Repina & Okuneva (1969) described *Khankaspis* from the Middle or Upper Cambrian. The body is known in *Angarocaris*, *Intejocaris*(?), *Chacharejocaris*, *Khankaspis* and *Schamanocaris*, in which it is of aglaspidid design. The chemical composition of the exoskeleton is unknown to us. Chernyshev (1945, p. 62) mentioned that the exoskeleton is 'shiny' in *Obrutschewia*. This glossiness, and the thickness and coarse sculpture seen in photographs, is evidence of mineralization of the exoskeleton in at least some forms, including *Angarocaris*, *Chacharejocaris*, *Girardevia*, *Obrutschewia* and *Schamanocaris*. Thus, both shape and mineralization indicate that the Eurasian

forms, or some of them, really are aglaspidids. Harry Mutvei (Swedish Museum of Natural History) has shown us mineralized fragments of Siberian aglaspidids.

Hesselbo (1989) assigned *Beckwithia* to the Aglaspididae. This would make the Beckwithiidae Raasch, 1939 a junior synonym of the Aglaspididae, a possibility that should await confirmation based on the exact count of body segments (which may be stable within the family) and data on the chemistry of the exoskeleton.

Indirect evidence indicates that strabopids and lemoneitids also had a mineralized exoskeleton (see discussion under the Order Strabopida, below). These forms, like the aglaspidids, have eleven body segments succeeded by a tail spine.

The aglaspidids used to be considered as xiphosurids. However, Bergström (1968) pointed out their distinctiveness and separated them as a subclass, the Aglaspida. At this time aglaspidid appendages were not known, and the critical factor for determining their possible affinities with chelicerates was therefore not at hand. Briggs *et al.* (1979) did not agree between themselves on the number of legs in the head, nor on the interpretation of the 1st appendage as an antenna or a chelicera. Hesselbo (1992) believes that there are four pairs of legs in the head. We think that the presence of a hypostome-like rostral doublure, as seen in *Lemoneites* (Flower 1968, Pl. 8:4, 13) and *Aglaspis* (Hesselbo 1992, Fig. 6), for functional reasons is incompatible with the development of chelicerae. Handling of food with chelicerae necessitates free access to the mouth from below and not only from behind. We therefore regard the 1st appendage as probably an antenna.

Briggs *et al.* (1979) and Hesselbo (1992) could not find evidence of an exopod in their material and discarded the evidence from *Khankaspis* (Repina & Okuneva 1969, e.g., Pl. 15:5). In our view, one may possibly question the affinities of *Khankaspis*, despite its similarity to North American aglaspidids, but the original illustrations clearly show the presence of trilobite-type lamellar setae. Furthermore, the lateral deflection of the legs and the position of the eyes on the head shield are typical of lamellipedian arthropods. The thick tail spine indicates that *Khankaspis* may be a member of the Strabopidae.

In summary, aglaspidids are typical lamellipedians, although unusual in having a mineralized exoskeleton.

Strabopids

Order Strabopida, n.ord.

Name. – From the Family Strabopidae.

Diagnosis. – Lamellipedians with rounded back (no trilobation), short head, sessile compound eyes of merostome type, and 11–12 body segments succeeded by a notably thick tail spine.

Families and genera include the Strabopidae Gerhardt, 1932 (=Paleomeridae Størmer, 1956), containing *Strabops* Beecher, 1901, *Neostrabops* Caster & Macke, 1951, and *Paleomerus* Størmer, 1955; and Lemoneitidae Flower, 1968, containing *Lemoneites* Flower, 1968. *Caryon* Barrande, 1872, has been suggested as a possible member of the Strabopidae (cf. Chlupáč 1963a, b). *Khankaspis* Repina & Okuneva, 1969, is another possible strabopid (see above).

This group is known from the Lower Cambrian to the Upper Ordovician. Although appendages are not known, these animals clearly fit well as merostomoids among the lamellipedians. In the *Treatise of Invertebrate Paleontology*, the strabopids are placed with the aglaspidids in the Xiphosura. The head tagma is quite obviously too short for a chelicerate. At most it comprises the antennal segment plus perhaps one or two leg segments, which excludes them from the Chelicerata. The strabopids show a general similarity in body shape with the aglaspidids (as expressed in Størmer's classification) and with *Sidneyia* and *Emeraldella*.

Flower (1968) compared Lemoneites with synziphosurid merostomes and with strabopids and placed it in the Aglaspidida, the taxon which at that time contained the strabopids. We find no reason to object to Flower's conclusion. Strabopids and lemoneitids have a semicircular head outline, eleven body tergites and a long and notably thick posterior element, the presumed telson. The original chemical composition of the exoskeleton is unknown. In strabopids, the exoskeleton is not preserved, but the 3-dimensional preservation indicates that it was sturdy and presumably calcareous. Phosphatic (but not calcareous) shells are preserved in the beds containing *Paleomerus*. In *Lemoneites* the exoskeleton is silicified; together with the thickness of the tergites, this indicates that it was originally calcareous.

Chelicerates

No chelicerates have been identified by us in the Chengjiang fauna. Although chelicerates differ from typical lamellipedians in lacking biramous appendages, lamellar setae, and antennae, and have preorally positioned chelicerae, many features point toward their close affinities with lamellipedians. It appears probable that they have both endopods and exopods, but on different segments; setae could have been lost on land; antennal glomeruli in the brain show that the ancestors had antennae; preoral appendages other than antennae are found also in the cheloniellid lamellipedians. Direct similarities between lamellipedians and merostome chelicerates include compound sessile eyes on the head shield, laterally directed legs and well-developed pleura. Aglaspidids and strabopids are so similar to merostomes that they have been mistaken for such (e.g., Størmer 1955; Ridley

1993, Fig. 19.12). Cambrian arthropods that may be chelicerates are few. They include the Lower or Middle Cambrian *Kodymirus*, which has eurypterid-like habitus, the 'correct' number of segments, merostome-type eyes, and non-mineralized integument (Chlupáč & Havlíček 1965), and the Lower Cambrian *Eolimulus*, of which only the prosoma is known (Bergström 1968). Bergström (1968) regarded them as chelicerates, a view not followed by others (Whittington 1979; Tollerton 1989; Selden 1993). However, the find of eurypterid-like legs in the prosoma of *Kodymirus* made Chlupáč (1995) regard this genus again as a eurypterid. The hypostome-like ventral plate is alien to later eurypterids but may be expected in early forms if these were derived from lamellipedians. *Eolimulus*, known from two prosomas, has typical xiphosuran features, and with *Kodymirus* accepted as a eurypterid its age is no longer any argument against a xiphosuran assignment.

Briggs & Collins (1988) suggested that *Sanctacaris uncata* is a primitive chelicerate based on (1) at least six pairs of head appendages, (2) head appendages raptorial as in eurypterids, (3) presence of cardiac lobe, (4) tagmosis of chelicerate type, (5) anus at base of telson, and (6) undivided telson with no associated appendages.

However, we do not agree that these arguments are valid. First, the number of head appendages is not known with certainty and, as also pointed out by Briggs & Collins (1988), there was already one case of parallel acquisition of that number, namely in *Emeraldella*. Herein we show that the lamellipedians *Xandarella* and *Almenia* also have a similar number. Secondly, adaptation to raptorial habits is no proof of relationship. Thirdly, the supposed cardiac lobe is the axial part with muscles, stomach, appendage bases, etc. It may just be that the thin pleura was bent out as a response to compaction. Fourthly, the tagmosis argument is partly the same as point one. The number of abdominal segments appear 'normal' for an arthropod and could be 11 just by chance. This, in fact, is different from the number in eurypterids, which is 12. Fifthly, several Burgess Shale arthropods appear to have had an anus in the same position, e.g., *Habelia*, *Molaria*, *Leanchoilia* and *Alalcomenaeus*. Sixthly, several Cambrian non-chelicerate arthropods have an undivided telson unassociated with paired appendages; in particular a very similar plate occurs in *Yohoia*. It lacks the keel seen in most merostome telsons.

Strong differentiation of the head appendages of *S. uncata* appears to be its greatest similarity with chelicerates. In detail, however, they differ (Briggs & Collins 1988). The legs are also pushed forwards to a degree not seen in any chelicerate, and there are no chelicerae. The eye position, at the margin of the head, is alien to the Chelicerata, but is more primitive and similar to the condition in crustacean-like arthropods and proschizoramians. The exopods, if correctly interpreted, are of the type found in proschizoramian arthropods and completely different from those in lamellipedians. The telson is most similar to that of proschizoramians such as *Alalcomenaeus* and *Yohoia*. The latter appears to have a strong tagmosis expressed in the legs, trunk exopods like those of *Sanctacaris*, eyes similarly positioned, and only two more trunk segments. We think that *Sanctacaris* lacks significant similarities with chelicerates and that it may be a proschizoramian, possibly with affinity to the megacheiran lineage.

Summary of morphological results

Segmentation

We are used to seeing arthropod segments as well-ordered packages, in which, e.g., each tergite behind the head corresponds to one pair of legs. There are exceptions among modern arthropods, such as the millipeds, in which tergites have fused two-and-two and therefore cover two pairs of legs, and the notostracans, in which the posterior legs outnumber the tergites. The former case is easily accepted as true segmentation, while the latter deviates from segmentation.

The Chengjiang arthropods in general demonstrate typical segmentation with marked serial similarity, such as would be expected from primitive arthropods. The arrangement with multisegment tergites posteriorly in *Xandarella* may be just a local modification in the tergum. Fusion of tergites is obvious in concilitergans, and a somewhat similar situation with only two large tergites is seen in *Naraoia*.

A very different situation occurs in *Fuxianhuia* and *Chengjiangocaris*, in which legs of a primitive design greatly outnumber the tergites. This may be a basic and original lack of correspondence between legs and tergites. Since legs and tergites are the only available evidence for possible segmentation, these discrepancies prevent conclusions on whether *Fuxianhuia* and *Chengjiangocaris* were truly segmented at all. In fact, if the disagreement is taken at its face value, it can only be taken as indication of pseudosegmentation, i.e. the state in which repetition of individual organs was not coordinated into truly segmental packages. Although the evidence is inconclusive, pseudosegmentation is widely distributed low down in the phylogenetic tree of animals (e.g., flatworms and aschelminths, and also in advanced forms such as molluscs), and therefore it may have preceded true segmentation (Bergström 1991). Logically, therefore, *Fuxianhuia* and *Chengjiangocaris* may be arthropods with clear organ repetition but yet without true segmentation, which is one of the most characteristic features of modern arthropods. However, we do not suggest that the two genera were not true arthropods. They are indeed arthropods in

having an exoskeleton divided into sclerites and in having segmented appendages.

Chen *et al.* (1995a) stated that *Fuxianhuia* has the same segmental composition in the head that- other arthropods show embryologically (and therefore have as adults). They regard this as evidence that *Fuxianhuia* is more primitive than other arthropods. They further describe the pseudosegmental type of lack of repetitive correspondence between body tergites and appendages and refer to it as segmentation.

Fuxianhuia (and *Chengjiangocaris*) is far from primitive in its tagmosis. The body is divided into no less than five distinct tagmata, namely the eye segment which may carry the carapace, the expanded tagma carrying specialized appendages and the mouth, the thorax with short tergites and perhaps three pairs of legs per tergite, the expanded opisthothorax with three pairs of legs for each tergite, expanding to about four pairs for the last tergite, and the cylindrical abdomen without legs (numbers are different in *Chengjiangocaris*). This is a very advanced condition, with more differentiation of tagmata than in modern arthropods. It seems to be an early offshoot, which had acquired its own specializations, but may also have preserved primitive features such as pseudosegmentation.

Tergum

The tergum is sclerotized in all arthropods we have identified. Except for the trilobites and aglaspidids, there is no case of undisputed mineralization. One specimen of *Fuxianhuia protensa* has a whitish tergum reminiscent of some phosphatic shells and skeletons, but this is an exception that is difficult to explain.

The tagmosis as reflected in the tergum varies considerably among the Chengjiang arthropods. There is always a head tagma, but its composition ranges from one to seven segments, in addition to the acron. The body may lack a further tagmosis, as in *Naraoia*, or it may consist of thorax and tail, as in *Sinoburius*, or of thorax and several 'tail' units, as in *Xandarella*, or of a short thorax and a long, subdivided abdomen, as in *Fuxianhuia*, or of just head and abdomen, as in *Jianfengia*. *Saperion* has no offset head but a single tergite covering the entire dorsum. There are invariably pleural folds, which may be narrow, as in more crustacean-like forms, or wide, as in *Naraoia* and most other Chengjiang lamellipedians.

Head

The Chengjiang arthropods present a wealth of information bearing on the formation of a head tagma. In probably all the Cambrian arthropods the (1st) antenna differs notably in its morphology from all the other appendages, even in cases when the latter are uniform in construction.

We believe that the (1st) antenna was originally uni-ramous. It is so in virtually all arthropods except for certain advanced crustaceans. It it possible that the antenna became specialized at a very early stage in the evolution of arthropods.

In many Cambrian arthropods, notably in the lamelli-pedians, head formation after an antennal segment was incorporated is merely a question of incorporating a specific number of additional segments into a tagma covered by a head shield. There is no indication that this was done successively, except perhaps in *Marrella*, with one post-antennal segment in the head, and the closely related *Mimetaster*, with two such segments. In other cases the incorporation of several segments may have been performed in a single step.

The shortest head shield that we know of appears to incorporate the presegmental acron and the antennal segment. Such a head is met with in *Sidneyia*. It is notable that in *Sidneyia* the paired eyes do not appear on the head shield but behind it, between the head shield and the first thoracic tergite. Still, if the condition is like that in modern arthropods, the eyes should be counted with the head and should in fact be connected with a more anterior part of the brain (the protocerebrum) than the antennae (tritocerebrum), which no doubt belong to the head if Bruton's (1981) count of segments and legs is correct.

The situation in *Fuxianhuia* is difficult to define. The eyes are associated with a separate tergite, but the head reasonably also includes the expanded portion behind. This portion of the body carries the mouth, and a transverse line indicates that the round portion may include two segments. In addition, the posterior expanded portion looks much like the succeeding tergites and may be a fourth segment included in the head. These structures were not noticed by Chen *et al.* (1995a). A head carrying two or more tergites is not unique among arthropods, as seen below. It may also be noted that there are extant mites carrying two prosomal tergites.

Most remarkably, there seems to be remnants of the *Sidneyia* 'stage' (head not including post-antennal segments) in some of the Chengjiang arthropods. In *Xandarella* there is a more comprehensive head, but there is still an open slit between anterior and posterior parts of the head shield. The paired eyes have a similar position as in *Sidneyia*, i.e. in the slit. The anterior plate may thus correspond to the head shield in *Sidneyia*. It is remarkable that the posterior head plate embraces as many as six leg segments. This makes the head probably identical in length with the arachnid prosoma, while the dwarfed segment at the boundary between head and thorax is a similarity with merostomes. *Xandarella* may be close to the origin of the chelicerates, but without additional similarities the mere similarity of segment numbers should not be relied on.

Another case of a possible suture between a primary head shield and an incorporated posterior plate is seen in *Kuamaia* and related genera, in which the anterior portion is separated by lines, possible sutures, into a median rostral plate and a pair of pararostral plates. It is interesting that the rostral plate has a comparatively large doublure, just as in the head shield of *Sidneyia*. A similarly large frontal doublure is seen in *Aglaspis* (Hesselbo 1992, Fig. 26:2).

Eyes

Where compound eyes are identified, they are present in either of several positions. First, they may be ventral and in such cases extend anteriorly under the anterior margin of the head shield, as in *Fuxianhuia*, *Leanchoilia*, *Isoxys* (see Shu *et al.* 1995) and *Waptia*. Among lamellipedians, this condition is seen in *Retifacies*, only that the eyes do not protrude anteriorly. Secondly, they can occur on the dorsal side of the head without being fused to the head shield. Thus, in the Burgess Shale *Sidneya* they extend laterally between the short head shield and the successive first thoracic tergite. In *Xandarella* the condition appears more advanced, since the eyes are higher up on the dorsal side, and since the tergite behind the eyes is formed by several segments and is partly fused with the anterior head shield. In the related *Cindarella*, the eyes are still ventral in position (Ramsköld *et al.* 1997). Ultimately, the eyes can be fused to the dorsal head shield. This is the case in *Kuamaia*, *Skioldia*, *Xandarella*, *Sinoburius* and most other lamellipedians with eyes.

Where sessile eyes are fused to the lamellipedian head, there are two distinct morphologies. In most cases, the eye surface is rounded, not particularly set off from the surroundings, and directed upwards. Trilobites, most strikingly the oldest ones, appear to be the only exception. In early trilobites the eye looks out horizontally through a long slit in the exoskeleton. The ocelli must have extended parallel with the tergal surface on the medial side of the slit, rather than at right angle to the surface as apparently is the case in other lamellipedians.

The morphological sequence probably indicates a corresponding evolutionary sequence of events. Thus, the compound eyes may have shifted position from ventral to dorsal and become successively incorporated into the head shield. However, the difference in organization between trilobites and other lamellipedians indicates that the former underwent the critical steps independently from the others.

Appendages

The Chengjiang arthropod specimens have been developed mechanically to expose many structures that were hitherto unknown or misunderstood. One significant new observation concerns the articulation between the two leg branches, another the structure of the 'trilobite' limb. The specificity of this kind of limb, and its importance for understanding early arthropod systematics, was realized by Størmer (1939, 1944), but his observations and conclusions have since been both neglected and dismissed (e.g., Briggs 1978b, p. 482: 'There is little justification for retaining the 'trilobitan limb' as a diagnostic character outside the trilobites'; Whittington 1979, p. 260: the 'trilobitan appendages' of Størmer as a 'premise is no longer valid'; Bruton 1981, p. 649: 'so-called similarity of the modified 'trilobite' appendage'; Bruton & Whittington 1983, p. 567: other authors had based their ideas on *Emeraldella* on its 'supposed trilobite-like appendages').

Endopods. – The primitive schizoramian appendage appears to have consisted of a stout endopod with many short podomeres of equal design, while the exopod may have been a simple rounded flap. In *Fuxianhuia* the number of podomeres is about 20, in *Chengjiangocaris* at least 17, in *Xandarella* 11–12 in addition to the terminal element. Similar multisegmented legs have previously been described from the Burgess Shale *Canadaspis*, which has 13 podomeres plus the terminal piece (Briggs 1978b), and from a Chinese *Tuzoia* (Shu 1990, Pl. 2:3). It is even found in the primitive uniramian euthycarcinoids (McNamara & Trewin 1993), where *Kottixerxes* had 24 segments and *Euthycarcinus* and *Kalbarria* about 12 segments. It is no longer possible to regard multisegmentation in the oldest arthropods as a secondary adaptation to flexibility needs, although such adaptation occurs in later arthropods (e.g., in cirripeds). It occurs in archaic types with thick, stubby legs, where it appears simply primitive. It is notable that the basis is serially similar to the podomeres of the endopod and also placed in direct continuation, and that the exopod is both different and placed at the side.

Exopods. – Based on two arguments, Whittington (1975, pp. 127, 132–133; 1980, pp. 188–189) concluded that the exopods of trilobites functioned as gills. One of these arguments is that the 'filaments' are wide enough to house blood vessels. However, also the endopod is wide enough to house blood vessels. This does not prove that it did, nor that is was a gill. More seriously, the fact that arthropods generally have a blastocoel with open circulation without fine blood vessels was not taken into consideration. Whittington has not given any argument other than the width of the 'filaments' (setae) for the presence of vessels in lamellipedians or in their appendages.

Whittington's other argument is that the ventral side of the body is virtually unknown (because of its softness). Therefore it would be less speculative to identify gills in preserved parts (Whittington 1980, p. 189). However, the tougher an integument is, the better is the fossilization

potential, and the smaller the chance that it may have functioned in a gill. A comparison with crustaceans, in which the underside of the carapace generally has the important 'gill' function, makes it very likely that much of the gaseous exchange was located on the underside of the pleural folds. Despite the flaws in Whittington's argument, his conclusion has been generally accepted for trilobites and trilobite-like arthropods, and the exopods are often not described, merely mentioned as 'gills' and 'gill filaments'.

The present study reveals a remarkable similarity between crustacean exopods with setae and trilobite-type exopods with 'filaments'. The 'filaments' are usually developed as demonstrably stiff lamellae with the flattened sides perpendicular to the surface of the exopod shaft (Størmer 1939, 1944; Bergström 1973, 1981, 1992 and others; Whittington 1985a). Where the preservation is good enough, a notable constriction is seen at the base of each lamella. This constriction marks the position of an articulation. A stiff outgrowth with such a basal articulation is by definition a seta. Thus, the 'filaments' of trilobite-type exopods are very long, flattened setae. Good examples of setal joints are seen in several genera, including *Emeraldella* (Bruton & Whittington 1983, Fig. 45, not mentioned in text), *Olenoides* (Whittington 1980, Pl. 22, not mentioned in text), and herein (e.g., *Naraoia*, *Retifacies*, and *Xandarella*).

In *Emeraldella brocki* there is less difference between setae of the proximal and distal exopod segments than in most other lamellipedians. Both types are flattened, but the distal ones are shorter than those on the proximal segment. The short distal spines appear to have one thick and one thin edge, and it is possible that the surfaces of the seta is inclined rather than perpendicular to the surface of the axis. These characters are more difficult to observe for the long proximal setae. Anyway, there is a possibility that the setae in *Emeraldella* are less derived than in other lamellipedians, and so they may indicate an origin for this type of seta.

The setae of the lamellipedian exopod form one or two rows and are remarkably evenly and tightly spaced. The even spacing is shared with the megacheirans. Regular rows occur in crustaceans (for instance in the thoracic legs of ostracodes), but in many cases crustacean setae are more irregularly arranged and of uneven size.

Tail

An elongate tail is very common. The wide-bodied *Retifacies* has a long, narrow, partly jointed tail. Similar tails, jointed or not, are known from *Sarotrocercus*, *Habelia*, *Molaria* and *Burgessia*. A sturdier, pointed tail spine occurs in *Jianfengia*. A pointed tail spine is also present in *Leanchoilia*, *Actaeus*, *Yohoia*, *Emeraldella*, aglaspidids, *Kodymirus*, and younger merostomes. Thus, most of the megacheiran arthropods have an elongate, pointed tail. An enlarged but bluntly terminating tail occurs in *Chengjiangocaris* and, outside the Chengjiang fauna, in *Sanctacaris* and *Paleomerus*, and a somewhat similar design ocurs in *Sidneyia*. Lamellipedians generally have an expanded pleura, while crustaceans and their kin tend to have a furca. Thus, the tail often gives a general hint of systematic affinity.

Intestine

The intestine can be preserved either as a thin dark band or as a three-dimensionally preserved yellowish structure. In the latter case, the filling is undoubtedly sediment. It is also clear that it is mud ingested by the live animal, since it fills the full length of the gut and in cases presents a detailed mould of the gut wall. The kind of preservation tends not to vary within species. Therefore, it most likely tells something of the feeding habits, the dark flat band being found in carnivores and scavengers, the mud-stuffed gut in sediment-eaters.

   The gut is most easily preserved and identified when stuffed with mud. Among arthropods of the Chengjiang fauna, a mud-stuffed gut is found in *Vetulicola*, *Fuxianhuia*, *Canadaspis laevigata*, *Leanchoilia illecebrosa*, *Chuandianella ovata* and various lamellipedians: *Naraoia longicaudata*, *Naraoia spinosa*, *Retifacies*, *Squamacula*, *Xandarella*, *Almenia* and *Sinoburius*. It may be noted that *Fuxianhuia* differs from the others in often having dark-coloured grains, presumably phosphatic, in its gut. This may indicate that it did not depend entirely on mud-feeding. Outside the Chengjiang fauna, a sediment-filled gut is reported from *Canadaspis*, *Perspicaris*, *Plenocaris*, *Molaria*, *Burgessia* and *Naraoia*. This was summarised by Briggs & Whittington (1985, p. 152). Without mineralogical analyses, however, these authors hesitate to conclude that the filling is mud. In the case of the Chengjiang animals, it is much clearer that the filling really consists of mud.

   In the Chengjiang material, a gut preserved without mud filling is seen in the proschizoramian *Fortiforceps* and the lamellipedian *Kuamaia*. Burgess Shale proschizoramians in which dark flat guts or squeezed out dark matter have been reported include *Sanctacaris* (Briggs & Collins 1988, p. 787), *Leanchoilia* (Bruton & Whittington 1983, p. 574) and *Yohoia* (Whittington 1974, pp. 12–13, Pl. IV:3–4, V:2, VIII:5, and IX:3). The same condition occurs in *Odaraia* (Briggs 1981) and in the lamellipedians *Tegopelte* (Whittington 1985a, p. 1261), *Emeraldella* (Bruton & Whittington 1983, p. 567; stain in Figs. 1–2) and *Sidneyia* (remains of 'ostracodes', hyoliths and small trilobites in gut, see Bruton 1981, pp. 548 and 644, repeated by Briggs & Whittington 1985, p. 151).

   It is much easier to recognize the sediment in the Chengjiang material than in the Burgess Shale. We therefore conclude that the three-dimensionally preserved guts are more or less filled with sediment and that these animals were sediment-eaters. Most of the others were probably carnivores and/or scavengers, as maintained by Bruton & Whittington (1983, p. 567), among others.

The Burgess Shale

The Burgess Shale has been the most famous fossil lagerstätte in the Cambrian for almost a century. Its importance largely depends on the fact that the shale provided the oldest known fauna with preservation of soft tissues and non-mineralized skeletons. Fame has been added through the impact of influential writers.

   Though the preservation of Burgess Shale is usually claimed to be marvellous, the situation is not so simple. The fossils are strongly flattened, and the shale is very hard and difficult to prepare, which makes interpretation of the fossils difficult. For instance, proximal portions of appendages are rarely seen or seen only as fragments. Walcott made numerous mistakes in his interpretations, and many more mistakes have been added during later studies. Although the Burgess Shale thus provided us with a welcome glimpse of life in the Cambrian, the interpretations are commonly problematic and have led us wrong in many cases. However, additional lagerstätten with other virtues and other problems are now being added to the Burgess Shale, which means new possibilities to interpret animals and structures by combining evidence from the different occurrences. The Burgess Shale will have a new chance to add valuable information, provided that the material can be studied again with an open mind (cf. Simonetta 1992).

   We think that most modern investigators of the Burgess Shale have paid too little attention to fundamental features of the limbs that were pointed out already by Leif Størmer (1939, 1944). Actually, the lateral deflection of the limbs (and the associated flattening of the body), the presence and unique morphology of exopod setae (usually referred to as 'gill filaments'), the possession of a hypostome, and the position of the compound eyes, are neglected characters which seem to be typical of a large group of arthropods, and which therefore must come in very low in the evolutionary tree.

   We note that the Cambridge arthropod school has repeatedly used a shared number of segments in the head as evidence of relationship (Briggs & Whittington 1981; Briggs 1983, 1990; Whittington 1977, 1985a; Briggs & Collins 1988). There are strong reasons to reject this idea: Stürmer & Bergström (1978, pp. 78–79) noted that 'even closely related forms may have different numbers of head segments and appendages', and Bruton & Whittington (1983, pp. 576–577) concluded that 'discussion on fossil arthropod relationships based on head segmentation ... appears to be largely irrelevant and, at best, speculative'.

Another reason for mistakes in the interpretation of the Burgess Shale fossils is a techno-mechanical approach to the interpretation, rather than a biological one. One example is *Anomalocaris* (Whittington & Briggs 1985; Briggs & Whittington 1987), which was supposed to lack legs (despite the obvious presence of genes for segmented limbs) and to swim waving skeletal plates like the fins of a flounder. No explanation was offered as to how the plates could have been flexed or where the driving muscles could have been attached to the sclerites. We appreciate the difficulties involved in the interpretation and realize that the authors must have been uneasy making it, since their reconstruction differs considerably from their description (see Bergström 1986, 1987; Dzik & Lendzion 1988, Hou *et al.* 1995). The reasons for this difference are stated to be 'difficulties of interpretation' and that they 'considered it prudent to omit the not understood details from the reconstruction' (Briggs & Whittington 1987). However, if a pattern of segmentation on the dorsal side is exchanged for a non-segmented pattern, it is a fundamental change of body design. Delle Cave & Simonetta (1991, p. 234, Fig. 31) did not make it easier by saying that they basically accept the reconstruction of Whittington & Briggs (1985), while presenting a redrawn version of the reconstruction by Bergström (1986) and Dzik & Lendzion (1988). On the whole, the wealth of different reconstructions of Burgess Shale animals is a good illustration of the difficulties in the intepretations.

With this background, it seems fair to say that the reconstructions based on Burgess Shale fossils have both led and misled us, from Charles D. Walcott onwards, in our attempts to understand life and evolution in the Cambrian. The Burgess Shale fossils are no doubt important for our knowledge, but the number of alternative reconstructions is a serious problem. This should be considered before speculations on systematics and evolutionary patterns lead too far.

The low relief and the poor light contrast with the matrix of the Burgess Shale fossils make photography difficult, and many of the published photographs lack the details they were certainly meant to show. Even if the direction of light is given, it is often difficult for the reader to know if a specimen is shown from above or from below. Our interpretation, therefore, is to some extent based on a study of specimens in Washington, DC, by one of us (JB).

Approaches to relationship problems

A quick survey of the literature is enough to persuade anybody that there is a considerable disagreement on the relationships between various Palaeozoic arthropods. Pub-lished phylogenetic trees vary greatly, also those produced by the same author.

Delle Cave & Simonetta (1991) recently discussed at length their views on morphology, taxonomy and relationships. Much of it is of a general character and does not reveal the thoughts behind their conclusions on relationships. Some premises given by them are outdated, such as the supposed presence of coxa and labrum in early arthropods (cf. Walossek & Müller 1990). This also makes the comparison between the positions of the exopod(ite) in crustaceans and trilobites (e.g., Delle Cave & Simonetta 1991, p. 196) out of date. Furthermore, the idea of a nauplius larva in trilobites (Delle Cave & Simonetta 1991, p. 197) should be seen in the light of the observation (Walossek & Müller 1990) that the nauplius larva was not yet developed even in the 'stem-lineage crustaceans', which were much closer related to the crusteaceans.

We have to look at the actual discussion on particular groups to understand the approach used by Delle Cave & Simonetta (1991 and earlier). In their tree, Table IV, crustaceans are split into three groups with separate non-crustacean origins: Notostraca with Kazacharthra are placed with Acercostraca and Burgessiida, all with a large univalved shield but with completely different limbs; Leptostraca, Archaeostraca, Conchostraca and Ostracoda are placed with non-crustacean forms, and all have a bivalved 'carapace'; some malacostracans are placed with Lipostraca and Anostraca, *Yohoia* and *Alalcomenaeus*, all sharing the absence of a bivalved carapace. The authors do not take into account the general understanding that crustaceans share a unique set of features, including coxa, labrum, nauplius larva with three pairs of specialized anterior limbs including two pairs of antennae, and exopod with setae directed towards the endopod (Walossek & Müller 1990). *Yohoia*, for instance, did not have a single one of these characters, still Delle Cave & Simonetta include it with the crustaceans. Similarly, the trilobite limbs have not been used for judgements of affinities, nor other lamellipedian characters (cf. Bergström 1992). We now have strong evidence that even groups formerly held together within the Ostracoda do not belong together, in part are not even crustaceans (Hou *et al.* 1996).

Delle Cave & Simonetta (1991, Fig. 9, p. 208) bring together 'emeraldellids', sharing only a posterior spine but being very different in other characters, including the morphology of the legs, as far as these are known.

We do not deny the possibility that some of the phylogenetic conclusions reached by Delle Cave & Simonetta (1991) may be correct. However, the uneven quality and high selectivity of their data make their results rather unconventional and unreliable.

Bousfield (1995) based his concept of early arthropod relationships on the idea that the mode of feeding developed from feeding with raptorial pre-oral appendages as seen in anomalocaridids. According to him, evolution went through a transitional stage to a stage with gnathobasic feeding limbs (Bousfield 1995, Fig. 7). In this stage he places a host of arthropods that lack gnathobases and raptorial appendages, instead being characterized by a high degree of primitive serial similarity. Bousfield's idea that the mode of feeding should be considered is splendid, but his logic in deriving primitive, unspecialized appendages from highly specialized ones is surprising. This polarization of appendage evolution appears to be a result of his wish to accept anomalocaridids as arthropods.

Briggs and colleagues (Briggs & Whittington 1981; Briggs 1983; Briggs & Fortey 1989; Briggs *et al.* 1992) have launched a whole series of evolutionary trees involving the Burgess Shale and some other Cambrian arthropods. Appreciating the difficulties, we note that their trees differ greatly between themselves. The absolutely basic difficulty in making a natural tree is to sort out the order of evolutionary events correctly. Their different trees represent a series of attempts to do this. In our view, they have generally placed the acquisition of a definite number of segments in the head far too early in the sequence of events. In this case we agree with Delle Cave & Simonetta (1991, p. 191), who state that 'there are no obvious phyletic affinities between genera having the same number of cephalized segments', and we further agree that genera with different numbers of head segments may be closely related with one another.

Other specialists on the Burgess Shale arthropods have used a one-character approach. Thus arthropods with antenna plus a supposed three pairs of appendages in the head have been determined as trilobites, whatever other characteristics they may have (Whittington 1977, 1985a). Arthropods with other numbers of appendages have been said to belong to other phyla. In another paper, Bruton & Whittington (1983, pp. 576–577) regarded this approach as irrelevant and speculative.

Our approach is to consider all available evidence and to combine it in various ways to see when the phylogenetic result looks reasonable. This is not to say that characters should be seen only in combination. Using single characters for phylogenetic conclusions is not in itself impermissible. Few would deny that the amniote egg is a sufficient character for distinguishing the Amniota. Most characters, however, are simpler and have evolved more than once. In any case, almost any characters can be misleading if we do not understand the order in which they have occurred. For example, if homothermy in tetrapods is given more 'weight' than skull-roof patterns, mammals may be classified with dinosaurs and birds rather than with mammal-like reptiles.

In evolution, characters are often added one by one. In order to sort out phylogenetic branches correctly, it is therefore very important to look at each character separately. When a character is placed in its proper position in a sequence of character acquisitions, it can be very powerful and decisive, even if the character is insignificant. Outside this position, it is usually misleading. It is this order that is important, not a general 'weight' of a character.

We believe that the number of segments have decreased in limbs and, at the very initiation of tagmosis, increased in heads. Thus, particular numbers have been acquired by parallel evolution, and the most parsimonious tree involving these characters may not be closest to the truth. Very specific modifications, such as the unique setal morphology in lamellipedians, are considered much more significant, particularly if they go along with other characters (Bergström 1992 and below). A similar suite of convincing modifications led to the crustaceans (Walossek & Müller 1990). It is then quite easy to recognize a lamellipedian or a true crustacean, once sufficient morphological data is available. *Canadaspis*, the alleged malacostracan crustacean, may serve as an example. It has some malacostracan-like features but lacks evidence of key crustacean characters such as nauplius larvae with three pairs of appendages, specializations of these appendages in the adult, labrum, coxa, and flagelliform exopod with setae directed inwards (Briggs 1992; Dahl 1992). The uniramous, supposed 2nd antenna (which is more similar to a 1st antenna), the absence of a coxa, and the blade-like exopod without setae are features that are alien to crustaceans, even though Whittington (1979, p. 257) considered *Canadaspis* to be similar in its limbs to Recent leptostracan crustaceans. The evidence forwarded by Walossek & Müller (1990) thus falsifies the interpretation of *Canadaspis* as a crustacean.

After sorting out groups such as the Crustacea and Lamellipedia, the remaining forms are much fewer and easier to overview. Comparisons between groups may suggest the order of evolutionary events. Unfortunately, many fossils without information on limb morphology are impossible to place systematically. Crustacean-type carapaces have evolved repeatedly and are *incertae sedis* if not associated with better evidence.

Evolutionary innovations are most clearly presented in diagrams using cladistic practice. As indicated above, however, the great amount of parallel evolution makes it dangerous to use any standard method for reconstruction of the phylogenetic tree. The attempt by Wilson (1992) illustrates the danger. For instance, in his attempts *Canadaspis* invariably comes out as a member of the Crustacea, despite its lack of the most fundamental crustacean characters.

Polarization of evolutionary changes

Segmental length of head shield. – There is a wide range in the length of head shields in early arthropods. The shield may not include any post-antennal segment, as in *Sid-*

neyia, or as many as seven such segments, as in *Xandarella*. In *Tegopelte* and *Saperion* the entire dorsum is covered by a single shield. However, there is no indication that the increase (or decrease) in length was the result of successive addition (or subtraction) through time. The differences between lineages in this respect appears to have been there virtually from the beginning. Furthermore, there is no correlation between length of head shield and any other factor (se for instance Briggs 1990). We conclude that the inclusion of post-antennal segments into the head shield was probably a matter of one step, and once it was finished, there was no further addition. The segmental length of the head shield therefore seems to be of no relevance to the discussion and discrimination of evolutionary lineages.

Labrum. – A labrum is present in most extant arthropods, except for parasites. Walossek & Müller (1990) has demonstrated that a labrum was absent in the oldest branches of 'stem-line crustaceans', and also Shu *et al.* (1995, p. 338) regard the absence of a labrum as plesiomorphic.

Morphology of ventral appendages. – Appendage features and structures showing morphological trends include the number of podomeres, the differentiation of podomeres, the morphology of the exopod, and the development of the portion proximal to the basis. The first two show a strong correlation: the more podomeres, the less differentiation of the individual podomeres. Thus in *Fuxianhuia*, with some 20 undifferentiated podomeres, the exopod is a simple rounded flap without any observable structure. The situation is very similar in *Canadaspis*, with close to 15 undifferentiated podomeres and a thin exopod flap with only faintly indicated lines of unknown significance. *Xandarella*, with 11–12 podomeres, exhibits the same lack of differentiation, whereas the exopod appears to be more advanced.

Extant arthropods tend to have fairly few podomeres. For instance, there are five in insects (including the composite tarsus). The large number found in cirripeds is obviously an adaptation to their particular way of straining food from the water and is not plesiomorphic for the larger group, the maxillopods, to which they belong.

Stance of ventral appendages. – It is useful to distinguish between pendant stance, in which the appendages of a pair are directed strictly ventrally, semi-pendant stance, in which they are directed ventrolaterally, and lateral stance, in which they are directed laterally, at least at the base. The fully pendant stance is confined to filter-feeding groups of crustaceans, and the lateral stance to a subgroup of lamellipedians. Thus, these two states are clearly within-group modifications, and the semi-pendant stance is the plesiomorphic condition. It is characteristic for most Cambrian arthropods, except for a majority of the lamellipedians.

Tagmosis. – It is useful to distinguish between coordination of segments affecting only the dorsum, coordination affecting only the ventral side, and that affecting the entire organization of a segment.

Tagmosis affecting primarily the dorsal side is seen in many Palaeozoic arthropods, for instance in *Naraoia*, *Kuamaia*, and in many (but not all) trilobites. There are also arthropods with a complete absence of dorsal differentiation into tagmata, notably *Tegopelte* and *Saperion*, but these are exceptions. As a rule, there is at least a distinction between head and body even in Lower Cambrian forms. The character therefore is not helpful in attempts to polarize morphological evolution.

Tagmosis affecting either only the ventral side or the entire segments exhibits a different pattern. The (1st) antennae appear to be present in all known arthropods, unless they are secondarily reduced, as in certain parasites. There is a notable general difference between the oldest arthropods and modern forms in the degree of tagmosis. Ventral tagmosis is poorly developed in many ancient arthropods. In many lamellipedians, for instance, all leg pairs are basically identical, and changes along the series are gradual rather than abrupt. In the comparatively well-known trilobites it can be seen how a certain degree of tagmosis is developed in some younger forms. Thus, for instance, there are proximal endites only in the cephalic legs in the Devonian *Phacops* (*Chotecops*) (Stürmer & Bergström 1973), and muscle scars in Ordovician and younger illaenids demonstrate that cephalic legs had a much more complicated musculature than those more posteriorly, and that pygidial legs were weaker and much more crowded than those of the thorax. Therefore, in trilobites we see a tendency with time to split a homogeneous series of legs into two and even three tagmata.

Some non-lamellipedians exhibit distinct tagmosis even in the Lower Cambrian. Thus, for instance, *Canadaspis* has the body divided into an anterior part with legs and a posterior portion devoid of legs. This is also the case in *Fuxianhuia* and *Chengjiangocaris*, which in addition have the leg-bearing portion divided into anterior and posterior series showing a difference in the crowding of the legs.

Whereas the degree of tagmosis is generally low in old Palaeozoic arthropods, this is not so in modern arthropods. Even in myriapods, where the legs are similar throughout, there is a clear-cut distinction between the limbs of the head and the legs of the body. We conclude that the absence of tagmosis is plesiomorphic (see also Shu *et al.* 1995, p. 338). Tagmosis also tends to become more inclusive with time, starting with the ventral appendages and ultimately incorporating the segments themselves. The condition in advanced insects is extreme, with three distinct tagmata, a head with antennae and mothparts, a thorax with locomotory appendages, and an abdomen without appendages but with a concentration of internal organs.

Head appendages. – The trend in the morphological evolution of head appendages is very clear. It starts with limbs not morphologically distinguishable from those further back, and ends with very specialized mouthparts. In practice, this can mean an accretion of segments into the functional head, as is well known from the crustaceans. Most of the Chengjiang arthropods lack specific mouthparts, whereas all extant arthropods except parasites have specialized mouthparts.

Mode of feeding. – Bousfield (1995) suggested that feeding with raptorial pre-oral appendages was typical of the first arthropods. However, this does not fit the actual evidence. A large number of the Early and Middle Cambrian schizoramian arthropods lacked specialized appendages for handling food, and we find their guts stuffed with mud. This indicates that they ingested mud directly with the mouth, which was directed ventrally. *Fortiforceps* is one of the Early Cambrian exceptions: it had strong grasping appendages, and in our specimens the gut lacks mud filling. Bousfield's conclusion is strongly contradicted by the circumstance that there were many arthropods (for instance all concilitergans, most xenopodans and many trilobites) that did not use mud-feeding but still lacked appendages specialized for grasping or manipulating food. (For the record, there is no known example of the opposite condition, namely arthropods with mud-filled guts and grasping appendages.) *Kuamaia lata*, one of the concilitergans, lacks mud filling and has no particular mouth appendages, but its endopods have strong medial spines, apparently for handling prey.

Number of limbs per segment. – *Fuxianhuia* and *Chengjiangocaris* have up to about four pairs of limbs per body segment. This is a unique condition in the Chengjiang material, since the condition in *Xandarella* and *Almenia* can be identified as the result of fusion of body segments, just as in extant diplopods and Carboniferous to Jurassic euthycarcinoids. Extant notostracan crustaceans have an excess number of abdominal limbs in comparison to the number of body segments. It is therefore not possible to recognize any evolutionary trend in this respect. There is a theoretical possibility that the condition in *Fuxianhuia* and *Chengjiangocaris* is a reminiscence from an early stage in which segmentation was not yet fully stable. However, it is obvious that these two arthropods have an advanced degree of tagmosis despite the primitive appearance of the limbs, and it may be at least as likely that multiplication is an aspect of this tagmosis, just as in notostracans.

Compound eyes. – Many of the Chengjiang arthropods have compound eyes. Except for the eyes of many lamellipedians, they have an anteroventral position. This condition is also typical for extant crustaceans. Exceptions

among crustaceans include notostracans, in which the eyes develop ventrally but migrate dorsally through the body, and isopods, which form a group high in the phylogenetic tree. A case for a dorsal origin can be made only if the condition in some lamellipedians is plesiomorphic, but this appears very unlikely. One reason for this is that in lamellipedians there is a range from entirely ventral eyes to dorsal eyes, and it is the dorsally positioned eyes that in some cases have left a 'migration trace' behind, notably in *Xandarella*. The anteroventral position is thus most likely plesiomorphic.

Pleura. – Many arthropods lack a pleural extension on the body segments. This is the case in, e.g., *Canadaspis*, *Burgessia*, *Marrella* and *Waptia*. In others, such as *Leanchoilia* and most lamellipedians, there are well-developed pleural folds. It is not immediately obvious which state is plesiomorphic. However, we tend to regard the lack of pleural folds as more primitive. One reason for this is that a more or less worm-like ancestor of arthropods presumably would have no pleural folds, since soft folds would serve no identifiable purpose. Another reason is that this polarity would fit best with the polarity of other characters. Pleural folds appear to have formed independently a number of times. Therefore the mere presence in two forms is no proof of relationship. However, the development of very wide and commonly fairly horizontal pleura in many lamellipedians seems to be specific to this group.

Conclusion

It is notable that several of the trends are correlated. Thus, the largest number of podomeres, the lowest degree of podomere differentiation, and the lowest degree of tagmosis is seen in the oldest arthropods. In fact, tagmosis is often absent behind the antennal segment, a condition never met with in extant arthropods. It is therefore clear that evolution has proceeded from a state with many similar legs with many similar podomeres towards a state with fewer legs and podomeres and with differentiation both between and within the legs. Similarly, evolution of body segments has proceeded from overall similarity to distinct tagmosis. One aspect of this is the successive differentiation of mouthparts in the head. Regarding the leg morphology, a simple non-segmented exopod flap attached laterally to the pediform endopod is found with the most plesiomorphic, multisegmented and undifferentiated endopods. Regarding segmental repetition, the general conclusion must be that evolution has gone from low towards high differentiation between serially repeated units. This conclusion receives strong support from the circumstance that several, perhaps all, of the character states regarded as plesiomorphic can be found in individual species, whereas advanced character states are much more scattered on different species and groups.

The compound eyes are, as a rule, anteroventral in position, except in some late and advanced crustacean groups, some lamellipedian groups, and chelicerates, where the evolutionary pathway from the ventral side can occasionally be traced. The anteroventral position is therefore regarded as the plesiomorphic condition.

The evolutionary trends being mapped, it is possible to deduce the morphology of the first common ancestor. This may not be the first arthropod, since the antennae already differed from the other limbs in being uniramous, and there was a tail. A pair of compound eyes was present anteroventrally. Behind the antennal segment was a series of body segments, all of roughly the same shape and each carrying pediform, multisegmented legs with a simple lateral exopod flap. Whether or not the last few segments carried legs is not important, but they presumably did. The dorsum was covered by tergites, each behind the antennal segment presumably covering only one segment.

Relationships and evolution

Outgroup comparison is a standard method of polarizing character states in phylogenetic studies. In the past, arthropods have generally been thought to have descended from annelids, and so it has been tempting to compare them with annelids. However, some molecular evidence indicates that annelids and arthropods are not closely related, and the anatomy and biology indicates that segmentation was brought about for different reasons and with in part different organ systems involved. If phyla related to arthropods exhibit fundamentally different body plans, we do not think that they can be used to polarize specific arthropod character states.

A few years ago, Budd (1993) regarded the Schizoramia to be a sister group of anomalocaridids, the latter comprising *Anomalocaris*, *Opabinia* and *Kerygmachela*. The first two of these had previously been brought together by Bergström (1986, 1987) and Dzik & Lendzion (1988). Later on (Budd 1996), he placed *Kerygmachela*, the aschelminth tardigrades, *Opabinia* and *Anomalocaris* as successive offshoots from the line leading to arthropods.

Previously, Bergström (1986, Fig. 3A) had suggested that *Opabinia* and *Anomalocaris* probably had sclerotized legs. According to Budd, both *Kerygmachela* and *Opabinia* had lobopod appendages under a series of tergal flaps extending from dorsal tergites (Budd 1993, Fig. 3b; 1996, Figs. 3C, 7). If Budd's interpretation of the structure in the two genera is correct, they are fundamentally different from anomalocaridids. The Chinese material demonstrates the presence in anomalocaridids of notably expanded paired ventral flaps carrying distally a comparatively small, clumsy leg branch (Hou *et al.* 1995). There

is a range from one species of anomalocaridids with irregular wrinkling of the lobopod-like leg branch to another with a well developed segmentation of the same branch (Hou *et al.* 1995, Figs. 9, 10, 16).

We regard it as possible that Budd is correct in regarding anomalocaridids as related to *Opabinia* and *Kerygmachela*. This cannot be, however, if his interpretation of the leg structure is correct.

We are sceptical regarding the suggested relationships of arthropods. There are strong reasons not to regard tardigrades as related to arthropods. Although tardigrades mimic arthropods, they are aschelminths (pseudocoelomates), whereas arthropods are coelomates. They are therefore widely apart in the phylogenetic tree. *Opabinia* and *Kerygmachela* are said to be related to arthropods because they have biramous appendages, but in Budd's reconstructions they are decidedly uniramous. Regarding anomalocaridids, the Chinese material indicates that the primary feature is a pair of very large ventral flaps, with a leg-like appendage, at first unsegmented, then segmented, being ultimately formed along the distal side. The Chinese material also indicates that in biramous arthropods it is the leg branch that is the primary structure, with a thin exopod being ultimately formed on the outer side (see further below). More or less lobopod-like appendages appear to have evolved in parallel in elasipodan echinoderms, myzostomid annelids, tardigrades, the anomalocaridid (?opabiniid) group, onychophorans, and arthropod progenitors, in some cases giving rise to segmented legs. This just demonstrates the extremely common occurrence of convergence in evolution. The mere presence of legs can never be taken as evidence of relationship. Fryer (1996), and earlier, e.g., Bergström (1978 on morphological grounds, 1994 on molecular grounds), regard it as likely that even typical arthropods are at least diphyletic.

The appendages of *Kerygmachela* and *Opabinia* appear to be too poorly preserved for any conclusion on possible similarities with the appendages of other animals. However, other similarities with anomalocaridids indicate that they may be related to them and that perhaps the appendages after all were similar.

When anomalocaridids are compared with schizoramians with primitive characters, such as *Fuxianhuia*, it appears that starting points were decidedly different also in structures other than the appendages (Hou *et al.* 1995). Anomalocaridids virtually lacked an external skeleton, the most distinctive feature of arthropods. The first arthropods had no distinct mouthparts and certain advanced groups had mouthparts consisting of modified appendages, whereas anomalocaridids, like many aschelminths, had a circle of 'scales'. The eyes, anteroventral in early arthropods, are dorsal in *Opabinia* and anomalocaridids. The 1st pair of appendages, characteristically forming uniramous antennae in arthropods, are developed as grasping appendages in *Kerygmachela*, *Opabinia* and anomalocaridids. Budd's phylogenetic diagram (1996, Fig. 9) involves a series of parallel losses of the antennae, although this is not indicated. We therefore agree with Fryer (1996) that it appears most likely that arthropods evolved directly from a worm-like origin, presumably more than once.

If we try to polarize character states in arthropods without the use of outgroups, specialization of segments and tagmosis in general must be considered more advanced than a state in which segments and podomeres are similar throughout. The segmentation of appendages lends itself most easily to a study of such trends. We know that in extant arthropods the legs (endopods) tend to be slender and to have knees, and the podomeres tend to be fairly few (e.g., six in hexapods, often less in crustaceans) and of distinctly uneven lengths. We also know that many Cambrian arthropods appear much less specialized. In arthropods such as *Leanchoilia*, *Sanctacaris*, *Agnostus*, *Naraoia*, *Tegopelte*, *Olenoides* and *Marrella* the podomeres were of more similar length throughout the endopod and there was no knee in the leg. In others, notably *Canadaspis*, *Fuxianhuia* and perhaps *Chengjiangocaris*, the situation was even more generalized, without virtually any sign of differentiation between podomeres, with a very high number of short podomeres (some 12–20) and with very thick legs. There is even sign that this is one end of the spectrum, the original morphology of the endopod. In both *Fuxianhuia* and *Canadaspis* the exopod is a simple flap devoid of setae; this may also be the primitive condition.

In the search for a most primitive arthropod, in the sense that it should preserve as many original characters as possible, *Fuxianhuia* is an interesting choice. It has a unique combination of characters. There is no indication of tagmosis in the leg series behind the large appendages in the head and no differentiation of individual podomeres. Each endopod is composed of a large number of identical podomeres, and the exopod is a simple flap devoid of setae. The tail tergite is a simple triangular plate, closely comparable to that in other Cambrian forms of primitive body design. The eye segment appears to have a tergite of its own. At the same time, however, there are more body tagmata than in any other arthropod: if eye segment, expanded part with mouth, prothorax, opisthothorax and abdomen are distinguished, there are no less than five tagmata. Therefore, *Fuxianhuia* appears to have preserved certain very original features but to have developed a most advanced tagmosis. In our attempt to estimate the primitiveness of Chengjiang arthropods, the genera *Canadaspis*, *Naraoia*, *Retifacies* and *Xandarella* also stand out as having a low number of apomorphic modifications. It appears to be the characters shared between these forms that indicate the morphology of a common ancestor.

Delle Cave & Simonetta (1991, p. 205) discussed the phylogenetic position of *Fuxianhuia*, but without any information on the ventral side their attempt was futile. The biramous legs place this animal among the Schizoramia. The exopods are clearly not lamellipedian in character. The coarse, multisegmented endopods have their counterpart only among some superficially crustacean-like forms, such as *Canadaspis*. However, *Fuxianhuia* is unique in having a consistent lack of coordination in its 'segmentation' – there are more pairs of legs than tergites. Since pseudomery or false segmentation is widely distributed among animal phyla, in particular among those that are presumed to have been branched off early, it is reasonable to suggest that arthropods evolved from a stock of pseudomeric animals (Bergström 1991). Pseudomery has often been called metamery or segmentation, but it differs in a fundamental respect: although organs are repeated, their numbers are not coordinated as in true segmentation. Repetition of tergites and legs obviously is not coordinated in *Fuxianhuia*, which therefore may possibly be described as pseudomeric (although we do not know about other organ systems). *Fuxianhuia* may represent a stage in which some basic arthropod characteristics were achieved, but before pseudomery had shifted into true metamery. This again indicates that *Fuxianhuia* may represent a very early offshoot, which may not really be attributable to the crustacean and trilobite–chelicerate main branches of the Schizoramia. If the interpretation of *Fuxianhuia* as pseudomeric (or having pseudomeric remnants) is correct, this means that segmentation was invented or at least made more perfect among the arthropods only after the invention of an exoskeleton. It also means that segmentation was derived independently in arthropods, onychophorans and annelids.

The morphology and position of the exopod in *Fuxianhuia* (and other primitive Cambrian arthropods) appear to falsify the suggestion by Emerson & Schram (1990, 1991) that the biramous appendage arose from fusion between appendages of two neighbouring segments.

At least four groups are discernible on the next level. The order of branching is uncertain. First, the lamellipedians (trilobitomorphs) are typically flattened and have exopods with lamellar setae. Most of them also have laterally deflected appendages and compound eyes penetrating the head shield. Secondly, the crustaceomorphs form a group which is usually characterized by setiferous, commonly multisegmented exopods and by a furca. Thirdly, the megacheiran arthropods are characterized by the presence of a large and branched 2nd antenna. Fourthly, there are *Sanctacaris*, *Habelia* and *Mollisonia*, which appear not to fit into any of the other groups. Ultimately, there are arthropods with mixed plesiomorphic and apomorphic features such as *Fuxianhuia* and *Canadaspis*.

In the latter two genera, the biramous appendage consists of a multisegmented endopod and a simple exopod flap without segmentation, spines or setae. Taking this as a starting point, one can easily imagine how evolution of the endopod proceeded towards slenderer shape and fewer and more specialized podomeres. Simultaneously, the exopod evolved in two directions. In one, it was split up in the margin, eventually forming marginal spines, as in the megacheirans. In the alternative direction, the exopod shaft appears to have become very slender and beset with setae along one margin. In this line, leading to crustacean-like arthropods and lamellipedians, the exopod became segmented. There is considerable similarity between supposedly primitive crustacean and lamellipedian exopods, indicating a shared origin of the specialization. The strong contrast between the two groups in leg posture is diminished by the intermediate posture in lamellipedians such as *Marrella* (Whittington 1971), *Mimetaster* (Stürmer & Bergström 1976) and *Naraoia* (herein). In *Marrella* and *Mimetaster* there is also a one-to-one correspondence of podomeres and setae just as in many Cambrian crustacean-like forms (e.g., *Martinssonia*, Müller & Walossek 1986). This indicates that lamellipedians arose from crustacean-like arthropods through modification of the setae, and later by the outward bend of the legs and displacement of the compound eyes over the lateral margin and up on the dorsum.

Key innovations separating more advanced lamellipedians from marrellomorphs include strong lateral deflection of the appendages, forming of longer exopod podomeres, and strong development of pleura. The separation of the exopod of *Emeraldella* into a proximal segment with long setae and a distal segment with short setae is also found in *Naraoia* and in some trilobites, including *Olenoides*. It is possible that this is the primitive condition for the lamellipedian exopod and that other morphologies are derived from it. Thus a similarity in this respect between lamellipedians can not so far be used as proof of close relationship, only of the retension of a primitive state. One evolutionary trend was to make the exopod flagelliform and increase the number of segments in it. Thus, in the trilobite *Triarthrus* the distal segment is kept, but the proximal segment is divided into narrow segments which are so short that each carries only two setae, and in the marrellomorphs (*Marrella*, *Mimetaster*) and possibly in *Vachonisia* there is one seta to each segment. Complications have led to two successive setal rows in some lamellipedians, including *Xandarella*, and in the trilobite *Cryptolithus*. A double row was reconstructed for *Sidneyia* (Bruton 1981, Fig. 107c), but the details are unknown.

There is a striking difference between lamellipedians and all other Cambrian arthropods in the position of the compound eyes. What we see is that even in the most primitive (i.e. small) heads, the eyes are either in front of or behind a head tergite. In the latter case, the eye became engulfed by the head tergite when it grew by accretion

(e.g., *Xandarella*). The primitive position of the eye in *Sidneyia* and *Xandarella*, and the different modes used by the eye to penetrate the tergite in trilobites on one hand and most other lamellipedians and merostomes on the other, tell us that the incorporation of the eyes was performed separately in different groups. This means that the head tagma of the primeval lamellipedian perhaps included the antennal segment but that leg segments were only incorporated in the separate subgroups. Thus, the number of segments in the head does not tell us much about affinities between groups. It is quite another thing that the number can be stable within a group once a certain tagmosis is acquired, such as in merostomes, arachnids, myriapods and insects.

To make the story even more complicated, *Retifacies* has its stalked eyes on the ventral side, like non-lamellipedians. This could be a secondary return to a primitive condition, but more likely is a genuine retension of this condition. If this is the case, we must be very careful to conclude that any particular Cambrian arthropod was blind because we find no dorsal eyes. In fact, *Naraoia*, *Saperion* and others may just have had their eyes on the ventral side, in the primitive way.

The eyes, therefore, tell us a lot about head formation, but unfortunately the implication is that there must have been much parallel evoution. Trilobites, for instance, appear to have acquired their number of head segments independently from all other lamellipedians, since their mode of incorporating the eyes seems unique.

On the whole, the lamellipedians cannot be sorted in any evolutionary order, but only in a number of groups which all share the basic lamellipedian characters and have added characters of their own. For instance, there is no reason to think that the head was formed through successive addition but rather through the initial choice of boundary between tagmata. The shortest head is found in *Sidneyia* and has only the antennal segment included; the longest head is found in the very disparate *Emeraldella* and *Cheloniellon*, both with antenna plus five segments. In a similar way, *Sidneyia* may demonstrate an original position of the compound eyes between two tergites, and the fusion of tergites into a head shield brought with it individual solutions. In several forms, the eye did not penetrate to the surface (e.g., *Emeraldella*, *Naraoia*, *Vachonisia*). In others, the eye kept its rounded shape and looked straight up through the tergite (e.g., aglaspidids, *Skioldia*, *Kuamaia*, *Mimetaster*). A third solution was to split the carapace to form a narrow vertical slit, through which a narrow band of ocelli could look out horizontally (earliest trilobites, possibly *Tegopelte*).

An expanded tergum is found in different groups and may also represent a trend of parallel evolution, 'caused' by the lateral deflection of the appendages and the development of the large exopod comb.

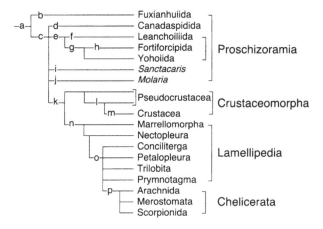

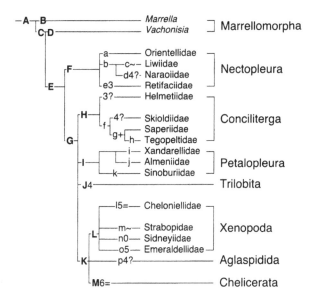

Fig. 87. Phylogeny and evolution of schizoramian arthropods, as interpreted herein. Letters denote evolutionary innovations. a, exoskeleton, anteroventral compound eyes, semi-pendent appendages consisting of multisegmented endopod and simple lateral exopod flap, telson flanked by pair of uropods, mud-eating habits; b, carapace and short thorax, complex tagmosis, particle food; c, segmentation improved to include correspondence between body segments and limb pairs; d, carapace, limbless abdomen; e, 2nd appendage transformed into 2nd antenna ('great appendage') with three flagellae, exopod fringed with setae, telson large; f, tail fan; g, 2nd appendage transformed into chelicera-like grasping organ, raptorial habits; h, long exopod setae, telson with marginal spines; i, grasping head appendages with reduced exopods, raptorial habits; j, long tail; k, slender multiarticulate exopod with setae directed away from endopod; l, setae directed towards endopod, 'coxal endite'; m, labrum, 2nd antenna and nauplius larva with three pairs of appendages, coxa, appendages pendent, abandonment of mud feeding; n, setae flattened (lamellar); o, lateral deflexion of appendages, flat exopod blade, well-developed pleura; p, 1st leg transformed into pre-oral chelicera, antenna lost, large prosoma comprising 5–6 pairs of appendages, appendages secondarily uniramous.

Fig. 88. Phylogeny and evolution of the Lamellipedia, as inferred herein. Numbers and symbols indicate head segments, where known; letters indicate other characters. *Upper-case letters*: A, trilobite-type lamellar setae, hypostome, possibly also protaspis larva; B, multisegmented exopod shaft, one seta per segment; C, more than one seta per segment; D, pleural folds fused along the body; E, pleural folds developed in individual segments; F, large pygidium (and eyes with primitive ventral position); G, laterally deflected legs, and compound eyes extending laterally between dorsal tergites; H, rostral plate small, pararostral plates, anteriorly positioned upward-looking eyes in head shield; I, wide pleural fold, formation of pygidial shield, tendency to include upward-looking eyes in head shield; J, trilobation, calcification, doublure, marginal (or facial) and circum-orbital suture, inclusion of sideward-looking eyes and four post-antennal leg pairs in head shield, eye ridge; K, upward-looking eyes included in head shield; L, tail/abdomen, and uropods?; M, 2nd pre-oral limb, loss of antenna. *Lower-case letters*: a, fusion between thoracic tergites; b, pygidium includes additional thoracic segments; c, thorax reduced to four segments; d, thorax lost by inclusion into pygidium; e, tail spine present, wide, sabre-shaped exopod setae; f, loss of pleural spines; g, complete fusion of tergum; h, effacement, i, posterior tergites covering more than one pair of legs, six or seven leg pairs in head tagma; j, loss of eyes; k, trilobation; l, 2nd pre-oral limb, head with two tergites; m, unknown; n, head formed by greatly enlarged acron and antennal segment, uropods and telson form tail fan; o, tail spine, and eyes lost(?); p, tail spine, and mineralization of exoskeleton. *Numbers and symbols for head segments*: 0, head without postantennal appendages; 1, a(ntenna) + one leg pair; 2, a+2; 3, a+3; 4, a+4; 5, a+5; 6, a+6; ~, not known; =, tagmosis judged from ventral side.

Lamellipedians are almost uniform in having one pair of preoral flagelliform antennae succeeded by a series of postoral legs. The single exception is the Devonian *Cheloniellon*, which has a second pair of preoral appendages. These seem to be grasping appendages. This fits in well with the absence of a protruding hypostome in this form. If *Cheloniellon* had kept the tail spine, added one or two segments to the head, and lost the antennae, it would have been a perfect ancestor to the chelicerates. It is morphologically intermediate between lamellipedians and chelicerates. The general position of the compound eye and the morphology of the legs of the head also represent similarities with the eurypterids. Of course it cannot be excluded that the similarity is due to parallel evolution.

Chelicerates were considered as a phylum of their own by Manton (1978). Many other authors believe that they somehow are related to, and derived from, early arthropods with bifid limbs (e.g., Størmer 1944; Bergström 1978, 1981, 1992; Lauterbach 1980; Fortey & Whittington 1989; Bruton 1991; Delle Cave & Simonetta 1991; etc.).

The disagreement on the details of this relationship sometimes appears to overshadow the general agreement.

Specifically, Simonetta & Delle Cave (1975) and Delle Cave & Simonetta (1991) derived chelicerates from the Emeraldellida (a wide concept in their opinion), Bergström (1978, 1981) from trilobitomorphs (i.e. lamellipedians), Lauterbach (1980) from trilobites, Bruton (1981) from some close relative of *Sidneyia*, and Bousfield (1995) from yohoiids. Delle Cave & Simonetta (1991) and Bous-

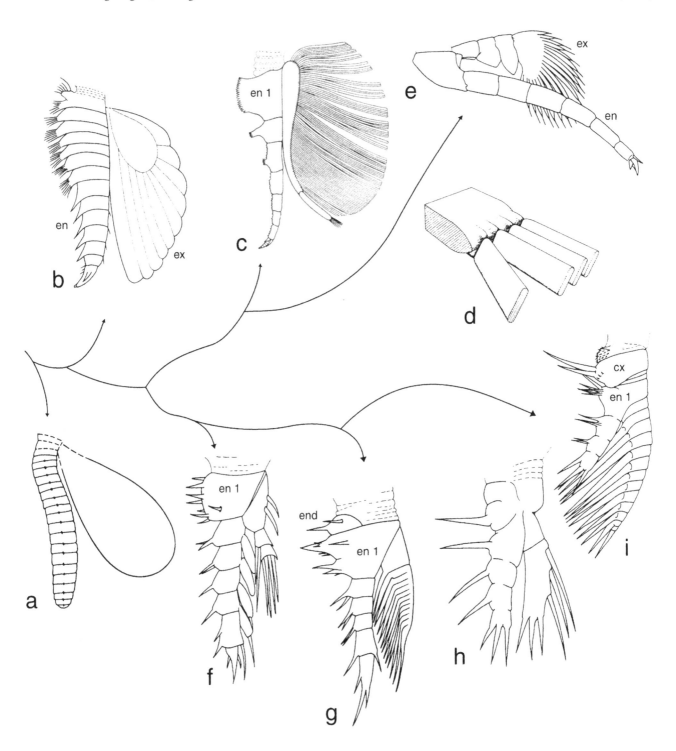

Fig. 89. Suggested evolution of the schizoramian appendage. a, stout multisegmented endopod representing major branch, with undifferentiated exopod in distinct lateral position (*Fuxianhuia*); b, like a, but with endites and with stiffening exopod structures (*Canadaspis*); c–i, setae developed on exopod, and soft cormus at base of limb; c–e, trilobite-like arthropods, setae with flattened cross section, with flat surface set at angle to exopod surface; c–d, (*Naraoia*); d, the insertion with proximal articulation typical of setae; e, (trilobite *Ceraurus*, modified from Størmer 1939); f–i, crustacean-like arthropods, setae rounded in cross section; f, 'pre-crustacean', without sign of coxal segment (4th appendage of *Agnostus*, redrawn from Walossek 1993); g, endite formed on soft cormus proximal to basis (2nd appendage of 'stemline crustacean' *Martinssonia*, redrawn from Walossek 1993); h–i, coxal segment fully developed proximal to basis (6rd and 3th appendages, i.e. maxilliped and mandibula of crustacean *Skara*, redrawn from Müller & Walossek 1985).

field (1995) believe that *Sanctacaris* is close to 'emeraldellids', and Briggs & Collins (1988) regard *Sanctacaris* to be a chelicerate. We hesitate to accept this interpretation of *Sanctacaris*.

In addition to *Sanctacaris* and aglaspidids, Cambrian arthropods that have been claimed to be chelicerates include *Kodymirus* and *Eolimulus* (Bergström 1968). *Kodymirus* has the appearance of a eurypterid, with kidney-shaped sessile eyes, the right number of segments, and a tail spine. Only the prosoma is known of *Eolimulus*, but it compares well with *Kodymirus* except that it has a marked median keel, which is a xiphosuran character. Delle Cave & Simonetta (1991, Fig. 10) accept *Kodymirus* as a 'post-emeraldellid' arthropod, which is their way of saying a chelicerate or related arthropod. This is perhaps as far as we can come without detailed information on the ventral morphology.

The evolutionary direction of modes of feeding was discussed above. Our conclusion is that evolution went from mud-eating without the use of appendages, to other kinds of feeding where appendages were utilized. Mud-eating seems to have been the habit in yunnanatans (*Fuxianhuia*), paracrustaceans (*Canadaspis*), pseudocrustaceans (e.g., *Chuandianella*), nectopleurans (e.g., *Naraoia, Retifacies*), and petalopleurans (e.g., *Xandarella*). It is also encountered in the leanchoiliids. All of these taxa lack true jaws and grasping appendages.

Mud-eating apparently was abandoned in crustaceans, concilitergans (e.g., *Helmetia, Kuamaia*), xenopodans (e.g., *Emeraldella, Sidneyia*), trilobites and chelicerates, and in the yohoiid and fortiforcipid megacheirans. The new mode of feeding was associated with strong modification of cephalic appendages in crustaceans, chelicerates and megacheirans, but not in others, which may have been rather unspecialized raptorians. The differences between the groups indicate that the new mode of feeding evolved independently. In yohoiids and fortiforcipids, for instance, the grasping appendage may have evolved from the leanchoiliid 2nd antenna through loss of the distal flagellae.

Summary of classification

It seems appropriate to place the arthropods of the Chengjiang fauna in systematic relation to other arthropods, particularly to those of the Burgess Shale and other Palaeozoic faunas. Below is a list including a selected number of arthropod groups ordered according to evolutionary classification (see Figs. 87–89 for summaries of phylogeny and evolution). A similar list was recently published by Bousfield (1995). It should be mentioned that many of the formal names in his list are unavailable. To mention one example, Bousfield lists the order Molariida referring it to Walcott, 1912. However, Walcott did not

mention any such order, but referred the genera *Habelia* to the family Aglaspidæ, Order Aglaspina (Walcott's spelling), of the Subclass Merostomata.

Following the evolutionary classification is a simplified attempt at a phylogenetic classification, in which the level-determined terminology presecribed by the International Code of Zoological Nomenclature is abandoned.

Evolutionary classification
Phylum Schizoramia
　Superclass Proschizoramia n. supercl.
　　Class Yunnanata n. cl.
　　　Order Fuxianhuiida Bousfield, 1995
　　　　Family Fuxianhuiidae n. fam.
　　　　Family Chengjiangocarididae n. fam.
　　Class Paracrustacea n. cl.
　　　Order Canadaspidida Novozhilov in Orlov, 1960
　　　　Canadaspididae Novozhilov in Orlov, 1960
　　　　Hymenocarididae Haeckel, 1896
　　　　　(? = Canadaspididae)
　　　　Perspicarididae Briggs, 1978
　　Class Megacheira n. cl.
　　　Order Leanchoiliida Størmer, 1944
　　　　Leanchoiliidae Raymond, 1935
　　　　Actaeidae Simonetta & Delle Cave, 1975
　　　　Alalcomenaeidae Simonetta & Delle Cave, 1975
　　　Order Yohoiida Simonetta & Delle Cave, 1975
　　　　Yohoiidae Henriksen, 1928
　　　Order Fortiforcipida n. ord.
　　　　Fortiforcipidae n. fam.
　　Class Sanctacaridea Bousfield, 1995
　　　Order Sanctacarida Bousfield, 1995
　?Classes uncertain
　　Order Acanthomeridiida, n. ord.
　　　Acanthomeridiidae n. fam.
　　Order Vetulicolida n. ord.
　　　Vetulicolidae n. fam.
　　Order not distinguished
　　　Habeliidae Simonetta & Delle Cave, 1975
　　Order not distinguished
　　　Molariidae Simonetta & Delle Cave, 1975
　　Order Tuzoiida Simonetta & Delle Cave, 1975
　　　Tuzoiidae Raymond, 1935
　　Order Bradoriida Raymond, 1935 ('thin chitino-
　　　calcareous-phosphatic shells')
　Superclass Crustaceomorpha Chernysheva, 1960
　　　('pan-crustaceans'),
　　Class Pseudocrustacea Størmer, 1944 ('stem-lineage
　　　crustaceans')
　　Order Agnostida Salter 1864
　　Order Waptiida Størmer, 1944
　　　Waptiidae Walcott, 1912 (*Waptia* Walcott, 1912,
　　　　Plenocaris Whittington 1974, *Chuandianella*
　　　　Hou & Bergström, 1991,)

Order Phosphatocopida Müller, 1964 'phosphatic
shells'
Order not distinguished
  Combinivalvula chengjiangensis Hou, 1987c
Class Crustacea Pennant, 1777
? Order Isoxyida Simonetta & Delle Cave, 1975 (as
Isoxyda)
  Isoxyidae Vogdes 1893 (Brooks & Caster, 1965)
Subclass Branchiopoda Latreille, 1817
Order Protocaridida Simonetta & Delle Cave, 1975
  Protocarididae Miller, 1889
Order Odaraiida Simonetta & Delle Cave, 1975
  Odaraiidae Simonetta & Delle Cave, 1975
Subclass Maxillopoda Dahl, 1956
Superclass Lamellipedia n.supercl.
Class Marrellomorpha (Beurlen, 1934) Størmer, 1944
Order Marrellida Raymond, 1935
  Marrellidae Walcott, 1912
Order Mimetasterida Beurlen, 1934
Class Artiopoda n.cl.
Subclass Nectopleura n.subcl.
Order Nectaspidida Raymond, 1920
  Orientellidae Repina & Okuneva, 1969
  Liwiidae Dzik & Lendzion, 1988
  Naraoiidae Walcott, 1912
Order Acercostraca Lehmann, 1955
  Vachonisia rogeri (Lehmann, 1955)
Order Halicyna Gall & Grauvogel, 1967
Order Retifaciida n.ord.
  Retifaciidae n.fam.
Subclass Conciliterga n.subcl.
Order Helmetiida Novozhilov, 1969
  Helmetiidae Simonetta & Delle Cave, 1975
  Skioldiidae n.fam.
  Tegopeltidae Simonetta & Delle Cave, 1975
  Saperiidae n.fam.
Subclass Trilobita Walch, 1771
Subclass Petalopleura n.subcl.
Order Xandarellida Chen *et al.* 1996
  Xandarellidae n.fam.
  Almeniidae n.fam.
Order Cheloniellida Broili, 1933 (emend.)
  Cheloniellidae Broili, 1933
Order Sinoburiida n ord.
  Sinoburiidae n.fam.
Subclass Xenopoda Raymond, 1935
Order Emeraldellida Størmer, 1944
  Emeraldellidae Raymond, 1935
Order Limulavida Walcott, 1911
  Sidneyiidae Walcott, 1911
Subclass Aglaspidida Bergström, 1968
Order Aglaspidida Walcott, 1911 (?=Beckwithiida
Raw, 1957)
  Aglaspididae Miller, 1887: several genera
  Lemoneitidae Flower, 1969

Order Strabopida n ordo
  Strabopidae Gerhardt, 1932 (=Paleomeridae
Størmer, 1955)
Superclass Chelicerata

Phylogenetic classification

Plesion Schizoramies
  Plesion Yunnanies
    Plesion Fuxianhuiides
    Plesion Chengjiangocarides
  Plesion Paracrustacees
    Plesion Canadaspides
    Plesion Perspicarides
  Plesion Megacheirides
    Plesion without name
      Plesion Leanchoiliides
      Plesion Actaeides
      Plesion Alalcomenaeides
    Plesion Yohoiides
    Plesion Fortiforcipides
  Plesion Sanctacarides
  Plesion without name
    Plesion Pan-crustacees
      Plesion Agnostides
      Plesion without name
        Plesion Waptiides
          Plesion Crustacees
            Plesion Branchiopodes
              Plesion Protocarides
              Plesion Odaraiides
            Plesion without name
              Plesion Maxillopodes
              Plesion Malacostraces
  Plesion Lamellipedes
    Plesion Marrellomorphes
      Plesion without name
        Plesion Marrellides
        Plesion Mimetasterides
    Plesion Artiopodes
      Plesion Nectopleurides
        Plesion Nectaspides
          Plesion Liwiides
          Plesion Naraoiides
          Plesion Orientellides
        Plesion Vachonisiides
        Plesion Halicynes
        Plesion Retifaciides
      Plesion without name
        Plesion Conciliterges
          Plesion Helmetiides
          Plesion Skioldiides
          Plesion Tegopeltides
          Plesion Saperiides
        Plesion Trilobites

Plesion Petalopleures
Plesion without name
  Plesion Xandarellides
  Plesion Almeniides
  Plesion Cheloniellides
  Plesion Sinoburiides
Plesion Xenopodes
  Plesion Emeraldellides
  Plesion Sidneyiides
Plesion without name
Plesion without name
  Plesion Aglaspidides
  Plesion Lemoneitides
  Plesion Strabopides
  Plesion Chelicerates

Acknowledgements. – We are most grateful for substantial grants and other support from Academia Sinica, the Royal Swedish Academy of Science, the Crafoord Foundation, the Swedish Natural Science Foundation (NFR), the Swedish Institute, and the Almén Fund, which have made it possible for us to arrange a close and fruitful cooperation over some years. The publication of this monograph was supported by the NFR. We appreciate very much the skilled and friendly support from Uno Samuelsson, who made our photographs, and Javier Herbozo, Lennart Andersson, and Björn Lindsten, who made the drawings. Dr. David Siveter improved the language and offered invaluable suggestions on the content. Last but not least, Dr. Graham Budd suggested improvements to the content.

References

Abele, L.G., Spears, T., Kim, W. & Applegate, M. 1992: Phylogeny of selected maxillopodan and other crustacean taxa based on 18S ribosomal nucleotide sequences: a preliminary analysis. *Acta Zoologica 73:5*, 373–382.

Andreeva, O.N. 1957: Novye nakhodki khlenistonogykh v vostochnoj Sibiri. [New discoveries on arthropods in eastern Siberia.] *Ezhegodnik vsesoyuznogo paleontologicheskogo obshchestva 16*, 236–243.

Andres, D. 1989: Phosphatisierte Fossilien aus dem unteren Ordoviz von Südschweden. *Berliner geowissenschaftliche Abhandlungen A 106*, 9–19.

Bergström, J. 1968: *Eolimulus*, a Lower Cambrian xiphosurid from Sweden. *Geologiska Föreningens i Stockholm Förhandlingar 90*, 489–503.

Bergström, J. 1973: Organization, life, and systematics of trilobites. *Fossils and Strata 2*. 69 pp.

Bergström, J. 1976a: Early arthropod morphology and relationships. *25th International Geological Congress, Abstracts*, p. 289. Sydney.

Bergström, J. 1976b: Lower Palaeozoic trace fossils from eastern Newfoundland. *Canadian Journal of Earth Science 13*, 1613–1633.

Bergström, J. 1978: Morphology of fossil arthropods as a guide to phylogenetic relationships. *In* Gupta, A.P. (ed.): *Arthropod Phylogeny*, 1–56. Van Nostrand Reinhold Co., New York.

Bergström, J. 1981: Morphology and systematics of early arthropods. *Abhandlungen des naturwissenschaftlichen Vereins Hamburg, NF 23* [for 1980], 7–42.

Bergström, J. 1986: *Opabinia* and *Anomalocaris*, unique Cambrian 'arthropods'. *Lethaia 19*, 241–246.

Bergström, J. 1987: The Cambrian *Opabinia* and *Anomalocaris*. *Lethaia 20*, 187–188.

Bergström, J. 1991: Metazoan evolution around the Precambrian–Cambrian transition. *In* Simonetta, A.M. & Conway Morris, S. (eds.): *The Early Evolution of Metazoan and the Significance of Problematic Taxa*, 23–34. Cambridge University Press, Cambridge.

Bergström, J. 1992: The oldest arthropods and the origin of the Crustacea. *Acta Zoologica 73:5*, 287–291.

Bergström, J. 1993: *Fuxianhuia* – possible implications for the origination and early evolution of arthropods. *In* Siverson, M. (ed.): *Lundadagarna i Historisk Geologi och Paleontologi 15–16 mars 1993, III, Abstracts*, 4. Lund University, Lund.

Bergström, J. 1994: Ideas on early animal evolution. *In* Bengtson, S. (ed.): *Early Life on Earth. Nobel Symposium No. 84*, 460–466. Columbia University Press, New York, N.Y.

Bergström, J. & Brassel, G. 1984: Legs in the trilobite *Rhenops* from the Lower Devonian Hunsrück Slate. *Lethaia 17*, 67–72.

Bergström, J., Briggs, D.E.G., Dahl, E., Rolfe, W.D.I. & Stürmer, W. 1987: *Nahecaris stuertzi*, a phyllocarid crustacean from the Devonian Hunsrück Slate. *Paläontologische Zeitschrift 61*, 273–298.

Boudreaux, H.B. 1979: *Arthropod phylogeny with special reference to insects.* 320 pp. Wiley, New York, N.Y.

Bousfield, E.L. 1995: A contribution to the natural classification of Lower and Middle Cambrian arthropods: food-gathering and feeding mechanisms. *Amphipacifica 2:1*, 3–33.

Briggs, D.E.G. 1976: The arthropod *Branchiocaris* n.gen., Middle Cambrian, Burgess Shale, British Columbia. *Geological Survey of Canada, Bulletin 264*, 1–29.

Briggs, D.E.G. 1978a: A new trilobite-like arthropod from the Lower Cambrian Kinzers Formation, Pennsylvania. *Journal of Paleontology 52:1*, 132–140.

Briggs, D.E.G. 1978b: The morphology, mode of life, and affinities of *Canadaspis perfecta* (Crustacea: Phyllocarida), Middle Cambrian, Burgess Shale, British Columbia. *Philosophical Transactions of the Royal Society of London B 281*, 439–487.

Briggs, D.E.G. 1981: The arthropod *Odaraia alata* Walcott, Middle Cambrian, Burgess Shale, British Columbia. *Philosophical Transactions of the Royal Society of London B 291*, 541–584.

Briggs, D.E.G. 1983: Affinities and early evolution of the Crustacea: the evidence of the Cambrian fossils. In Schram, F.R. (ed.): *Crustacean Phylogeny*, 1–22. Balkema, Rotterdam.

Briggs, D.E.G. 1990: Early arthropods: dampening the Cambrian explosion. *Arthropod Paleobiology. Short Courses in Paleontology 3*, 24–43. Paleontological Society, University of Tennessee, Knoxville, Tenn.

Briggs, D.E.G. 1992: Phylogenetic significance of the Burgess Shale crustacean *Canadaspis*. *Acta Zoologica 73*, 293–300.

Briggs, D.E.G., Bruton, D.L., Whittington, H.B. 1979: Appendages of the arthropod *Aglaspis spinifer* (Upper Cambrian, Wisconsin) and their significance. *Palaeontology 22*, 167–180.

Briggs, D.E.G. & Collins, D. 1988: A Middle Cambrian chelicerate from Mount Stephen, British Columbia. *Palaeontology 31*, 779–798.

Briggs, D.E.G. & Fortey, R.A. 1989: The early radiation and relationships of major arthropod groups. *Science 246*, 241–243.

Briggs, D.E.G., Fortey, R.A. & Wills, M.A. 1992: Morphological disparity in the Cambrian. *Science 256*, 1670–1673.

Briggs, D.E.G. & Whittington, H.B. 1981: Relationships of arthropods from the Burgess Shale and other Cambrian sequences. *In* Taylor, M.E. (ed.): *Short Papers for the Second International Symposium on the Cambrian System*, 38–41. *US Department of the Interior, Geological Survey, Open-File Report 81-743*.

Briggs, D.E.G. & Whittington, H.B. 1985: Modes of life of arthropods from the Burgess Shale, British Columbia. *Transactions of the Royal Society of Edinburgh 76*, 149–160.

Briggs, D.E.G. & Whittington, H.B. 1987: The affinities of the Cambrian animals *Anomalocaris* and *Opabinia*. *Lethaia 20*, 185–186.

Broili, F. 1932: Ein neuer Crustacee aus dem rheinischen Unterdevon. *Sitzungsbericht der bayerschen Akademie der Wissenschaften 1932*, 27–38.

Broili, F. 1933: Ein zweites Exemplar von *Cheloniellon*. *Sitzungsbericht der bayerschen Akademie der Wissenschaften 1933*, 11–32.

Brooks, H.K. & Caster, K.E. 1956: *Pseudoarctolepis sharpi*, n.gen., n.sp. (Phyllocarida), from the Wheeler Shale (Middle Cambrian) of Utah. *Journal of Paleontology 30*, 9–14.

Bruton, D.L. 1981: The arthropod *Sidneyia inexpectans*, Middle Cambrian, Burgess Shale, British Columbia. *Philosophical Transactions of the Royal Society of London B 295*, 619–656.

Bruton, D.L. & Whittington, H.B. 1983: *Emeraldella* and *Leanchoilia*, two arthropods from the Burgess Shale, Middle Cambrian, British Columbia. *Philosophical Transactions of the Royal Society of London B 300*, 553–585.

Budd, G. 1993: A Cambrian gilled lobopod from Greenland. *Nature 364*, 19 August 1993, 709–711.

Butterfield, N.J. 1994: Burgess Shale-type fossils from a Lower Cambrian shallow-shelf sequence in northwestern Canada. *Nature 369*, 477–479.

Charig, A.J. 1990: Evolutionary systematics. *In* Briggs, D.E.G. & Crowther, P.R. (eds.): *Palaeobiology. A Synthesis*, 434–437. Blackwell, Oxford.

Chen J.-y., Bergström, J., Lindström, J. & Hou X.-g. 1991: Fossilized soft-bodied fauna. *National Geographic Research & Exploration 7:1*, 8–19.

Chen J.-y., Edgecombe, G.D., Ramsköld, L. & Zhou G.-q. 1995a: Head segmentation in Early Cambrian *Fuxianhuia*: implications for arthropod evolution. *Science 268*, 1339–1343.

Chen J.-y., Zhou G.-q. & Ramsköld, L. 1995b: A new Early Cambrian onychophoran-like animal, *Paucipodia* gen. nov., from the Chengjiang fauna, China. *Transactions of the Royal Society of Edinburgh 85*, 275–282.

Chen J.-y., Zhou G.-q., Zhu, M.-y. & Yeh, K.-y. 1996: *The Chengjiang Biota. A Unique Window of the Cambrian Explosion.* 222 pp. The National Museum of Natural Science, Taichung, Taiwan. (In Chinese.)

Chernyshev (Tchernychev), B.I. 1945: On *Obrutschewia* Tschern. and other Arthropoda from the Angara River (Siberia). *Ezhegodnik vsesoyuznogo paleontologicheskogo obshchestva 12*, 60–68.

Chernyshev, B.I. 1953: Novye chlenistonogie s. r. Angary. *Ezhegodnik vsesoyuznogo paleontologicheskogo obshchestva 14*, 106–1126.

Chlupáč, I. 1963a: Report on the merostomes from the Ordovician of central Bohemia. *Vestnik Ustredniho Ustav geol. 38*, 399–403.

Chlupáč, I. 1963b: Xiphosuran merostomes from Bohemian Ordovician. *Sbornik geologickych ved, P5*, 7–38.

Chlupáč, I. 1995: Lower Cambrian arthropods from the Paseky Shale (Barrandian area, Czech Republic). *Journal of the Czech Geological Society 40:4*, 9–36.

Chlupáč, I. & Havlíček, V. 1965: *Kodymirus* n.g., a new aglaspid merostome of the Cambrian of Bohemia. *Sbornik geologickych ved, P 6*, 7–20.0

Cisne, J.L. 1975: Anatomy of *Triarthrus* and the relationships of the Trilobita. *Fossils and Strata 4*, 45–63.

Conway Morris, S. 1977: A new metazoan from the Cambrian Burgess Shale, British Columbia. *Palaeontology 20*, 623–640.

Conway Morris, S. 1979: Views on an evolutionary question. *Science 204*, 1299.

Conway Morris, S. 1993: Ediacaran-like fossils in Cambrian Burgess Shale-type faunas of North America. *Palaeontology 36*, 593–635.

Dahl, E. 1984: The subclass Phyllocarida (Crustacea) and the status of some early fossils: a neontologist's view. *Videnskabelige Meddelelser fra dansk naturhistorisk Forening 145*, 61–76.

Dahl, E. 1987: Malacostraca maltreated – the case of the Phyllocarida. *Journal of Crustacean Biology 7*, 721–726.

Delle Cave, L. & Simonetta, A. M. 1991: Early Palaeozoic arthropods and the problems of arthropod phylogeny; with some notes on taxa of doubtful affinities. *In* Simonetta, A.M. & Conway Morris, S. (eds.): *The Early Evolution of Metazoan and the Significance of Problematic Taxa*, 189–244. Cambridge University Press, Cambridge.

Dzik, J. & Lendzion, K. 1988: The oldest arthropods of the East European Platform. *Lethaia 21*, 29–38.

Emerson, M.J. & Schram, F.R. 1990: The origin of crustacean biramous appendages and the evolution of Arthropoda. *Science 250*, 667–669.

Emerson, M.J. & Schram, F.R. 1991: Remipedia. Part 2. Paleontology. *Proceedings of the San Diego Society of Natural History 1991:7*, 1–52.

Erdtmann, B.-D., Steiner, M. & Siegmund, H. 1994: Entwicklungsgeschichte des höheren Lebens an der Wende vom Präkambrium zum Kambrium. *Aufschluss 45*, 26–35. Heidelberg.

Flower, R.H. 1968: Merostomes from the Cassinian portion of the El Paso Group. *New Mexico Bureau of Mines and Mineral Resources, Memoir 22*, 35–44.

Fortey, R.A. & Theron, J.N. 1995: A new Ordovician arthropod, *Soomaspis*, and the agnostid problem. *Palaeontology 37:4*, 841–861.

Fortey, R.A. & Whittington, H.B. 1989: The Trilobita as a natural group. *Historical Biology 2*, 125–138.

Fryer, G. 1985: Structure and habits of living branchiopod crustaceans and their bearing on the interpretation of fossil forms. *Transactions of the Royal Society of Edinburgh 76*, 103–113.

Fryer, G. 1996: Reflections on arthropod evolution. *Biological Journal of the Linnean Society 58*, 1–55.

Gould, S.J. 1989: *Wonderful Life. The Burgess Shale and the Nature of History.* 347 pp. Norton, New York, N.Y.

Hammann, W., Laske, R. & Pillola, G.L. 1990: *Tariccoia arrusensis* n.g. n.sp., an unusual trilobite-like arthropod. Rediscovery of the 'phyllocarid' beds of Taricco (1922) in the Ordovician 'Puddinga' sequence of Sardinia. *Bolletino della Societa Paleontologica Italiana 29:2*, 163–178.

Henriksen, K.L. 1926: The segmentation of the trilobite's head. *Meddelelser fra Dansk geologisk Forening 7*, 1–32.

Hesselbo, G.P. 1989: The aglaspidid arthropod *Beckwithia* from the Cambrian of Utah and Wisconsin. *Journal of Paleontology 63*, 636–642.

Hesselbo, G.P. 1992: Aglaspidida (Arthropoda) from the Upper Cambrian of Wisconsin. *Journal of Paleontology 66*, 885–923.

Heymons, R. 1901: Die Entwicklungsgeschichte der Scolopender. *Zoologica 33*, 1–244. Stuttgart.

Ho C. S. (He Chun-cun) 1942: Phosphate deposits of Tungshan, Chengjiang, Yunnan. *Bulletin of the Geological Survey of China 35*, 97–106 (in Chinese), 41–43 (in English). Pepei, Chungking.

Hou X.-g. 1987a: Two new arthropods from Lower Cambrian, Chengjiang, eastern Yunnan. *Acta Palaeontologica Sinica 26:3*, 236–256. (In Chinese, with English summary.)

Hou X.-g. 1987b: Three new large arthropods from Lower Cambrian, Chengjiang, eastern Yunnan. *Acta Palaeontologica Sinica 26:3*, 272–285. (In Chinese, with English summary.)

Hou X.-g. 1987c: Early Cambrian large bivalved arthropods from Chengjiang, eastern Yunnan. *Acta Palaeontologica Sinica 26:3*, 286–298. (In Chinese, with English summary.)

Hou X.-g. 1987d: Oldest Cambrian bradoriids from eastern Yunnan. In: *Stratigraphy and Palaeontology of Systemic Boundaries in China. Precambrian–Cambrian Boundary 1*, 535–547. Nanjing University Publishing House, Nanjing.

[Hou X.-g. 1997: Bradoriid arthropods from the Lower Cambrian of Southwest China. 104 pp. Unpublished Ph.D. thesis, Uppsala University.]

Hou X.-g. & Bergström, J. 1991: The arthropods of the Lower Cambrian Chengjiang fauna, with relationships and evolutionary significance. *In* Simonetta, A.M. & Conway Morris, S. (eds.): *The Early Evolution of Metazoan and the Significance of Problematic Taxa*, 179–187. Cambridge University Press, Cambridge.

Hou X.-g. & Bergström, J. 1995: Cambrian lobopodians – ancestors of extant onychophorans? *Zoological Journal of the Linnean Society 114*, 3–19.

Hou X.-g., Bergström, J. & Ahlberg, P. 1995: *Anomalocaris* and other large animals in the Lower Cambrian Chengjiang fauna of southwest China. *Geologiska Föreningens i Stockholm Förhandlingar 117*, 163–183.

Hou X.-g., Chen J.-y., Lu H.-z. 1989: Early Cambrian new arthropods from Chengjiang, Yunnan. *Acta Palaeontologica Sinica 28:1*, 42–57. (In Chinese, with English summary.)

Hou X.-g., Ramsköld, L. & Bergström, J. 1991: Composition and preservation of the Chengjiang fauna – a Lower Cambrian soft-bodied biota. *Zoologica Scripta 20*, 395–411.

Hou X.-g., Siveter, D.J., Williams, M., Walossek, D. & Bergström, J. 1996: An early Cambrian bradoriid arthropod from China with preserved appendages: its bearing on the origin of the Ostracoda. *Philosophical Transactions of the Royal Society of London B 351*, 1131–1145.

Hou X.-g. & Sun W.-g. 1988: Discovery of Chengjiang fauna at Meichucun, Jinning, Yunnan. *Acta Palaeontologica Sinica 27:1*, 1–12. (In Chinese, with English summary.)

Huo S.-c. 1956: Brief notes on Lower Cambrian Archaeostraca from Shensi and Yunnan. *Acta Palaeontologia Sinica 4:3*, 425–445. (In Chinese, with English summary.)

Huo S.-c. & Shu D.-g. 1985: *Cambrian Bradoriida of South China.* 251pp. Northwest University Press, Xian, Shaanxi. (In Chinese, with English summary.)

Huo, S.-c., Shu D.-g. & Cui Z.-l. 1991: *Cambrian Bradoriida of China.* 249pp. Geological Publishing House, Beijing. (In Chinese, with English summary.)

Hughes, C.P. 1975: Redescription of *Burgessia bella* from the Middle Cambrian Burgess Shale, British Columbia. *Fossils & Strata 4*, 415–436.

Jiang Z.-w. 1982: Ostracoda. In Luo H.-l. *et al.* (eds.): *The Sinian–Cambrian boundary in Eastern Yunnan, China.* Yunnan People's Press, Kunming, 211–215. (In Chinese, with English summary.)

Lauterbach, K.-E. 1980: Schlüsselereignisse in der Evolution des Grundplans der Arachnata (Arthropoda). *Abhandlungen des Naturwissenschaftlichen Vereins in Hamburg 23*, 163–327.

Li (Lee) Y.-w. 1975: On the Cambrian Ostracoda with new materials from Sichuan, Yunnan and Southern Shaanxi, China. *Professional Papers of Stratigraphy and Palaeontology 2*, p. 37–72. Geological Publishing House, Beijing. (In Chinese.)

Lu Y.-h. 1941: Lower Cambrian stratigraphy and trilobite fauna of Kunming, Yunnan. *Bulletin of the Geological Society of China 21:1*, 71–90.

Lu Y.-h., Yu Ch.-m. & Chen P.-j. 1981: Invertebrate paleontology in China (1949–1979). *Geological Society of America, Special Paper 187*, 3–8.

Mansuy, H. 1912: Etude géologique du Yunnan oriental. Part 2. Paléontologie. *Mémoires du Service géologique de l'Indochine 1:2*, 1–146.

Manton, S.M. 1972: The evolution of arthropodan locomotory mechanisms. Part 10. Locomotory habits, morphology and evolution of the hexapod classes. *Journal of the Linnaean Society (Zoology) 51*, 203–400.

Manton, S.M. 1978 [date of imprint 1997]: *The Arthropoda.* 527 pp. Clarendon Press, Oxford.

McNamara, K.J. & Trewin, N.H. 1993: A euthycarcinoid arthropod from the Silurian of western Australia. *Palaeontology 36:2*, 319–335.

Moczydłowska, M. & Vidal, G. 1988: How old is the Tommotian? *Geology 16*, 166–168.

Moore, R.C. 1959: *Treatise on Invertebrate Paleontology, O, Arthropoda 1.* 560 pp. Geological Society of America and University of Kansas Press, Boulder, Colo.

Moore, R.C. & McCormick, L. 1969: General features of Crustacea. *In* Moore, R.C. (ed.): *Treatise on Invertebrate Paleontology, R, Arthropoda 4*, R57–R120. Geological Society of America and University of Kansas Press, Lawrence, Kansas.

Müller, K.J. 1981: Arthropods with phosphatized soft parts from the Upper Cambrian 'Orsten' of Sweden. *In* Taylor, M.E. (ed.): *Short Papers for the Second International Symposium on the Cambrian System*, 147–151. *US Department of the Interior, Geological Survey, Open-File Report 81-743.*

Müller, K.J. 1982: *Hesslandona unisulcata* sp.nov. with phosphatised appendages from Upper Cambrian 'Orsten' of Sweden. *In* Bate, R.H., Robinson, E. & Sheppard, L.M. (eds.): Fossil and Recent ostracods. *The British Micropalaeontological Society 1982*, 276–306.

Müller, K.J. 1983: Crustacea with preserved soft parts from the Upper Cambrian of Sweden. *Lethaia 16*, 93–109.

Müller, K. & Walossek, D. 1986: *Martinssonia elongata* gen. et sp. n., a crustacean-like euarthropod from the Upper Cambrian 'Orsten' of Sweden. *Zoologica Scripta 15*, 73–92.

Müller, K.J. & Walossek, D. 1987: Morphology, ontogeny, and life habit of *Agnostus pisiformis* from the Upper Cambrian of Sweden. *Fossils and Strata 19*, 1–124.

Neville, A.C. 1975: *Biology of the Arthropod Cuticle*, 448 pp. Springer, Berlin.

Palacios, T. & Vidal, G. 1992: Lower Cambrian acritarchs from northern Spain: the Precambrian–Cambrian boundary and biostratigraphic implications. *Geological Magazine 129*, 421–436.

Ramsköld, L. & Edgecombe, G.D. 1991: Trilobite monophyly revisited. *Historical Biology 4*, 267–283.

Ramsköld, L., Chen, J.y., Edgecombe, G.D. & Zhou, G.-q. 1996: Preservational folds simulating tergite junctions in tegopeltid and naraoid arthropods. *Lethaia 29*, 15–20.

Ramsköld, L., Chen, J.y., Edgecombe, G.D. & Zhou, G.-q. 1997: *Cindarella* and the arachnate clade Xandarellida (Arthropoda, Early Cambrian) from China. *Transactions of the Royal Society of Edinburgh: Earth Sciences 88*, 19–38.

Ramsköld, L. & Edgecombe, G.D. 1996: Trilobite appendage structure – *Eoredlichia* reconsidered. *Alcheringa 20*, 269–276.

Repina, L.N. & Okuneva, O.G. 1969: [Cambrian arthropods of the Maritime territory.] Kembrijskie chlenistonogie Primor'ya. *Paleontologicheskij Zhurnal 1969*, 106–114. (In Russian; English translation published by American Geological Institute.)

Ridley, M. 1993: *Evolution.* 670 pp. Blackwell, Boston, Mass.

Robison, R.A. 1990: Earliest-known uniramous arthropod. *Nature 343*, 163–164.

Robison, R.A. & Richards, B.C. 1981: Larger bivalved arthropods from the Middle Cambrian of Utah. *The University of Kansas Paleontological Institute, Paleontological Contributions, Paper 106*, 1–19.

Rolfe, W.D.I. 1969: Phyllocarida. *In* Moore, R.C. (ed.): *Treatise on Invertebrate Paleontology, R, Arthropoda 4*, R269–331. Geological Society of America and University of Kansas Press, Lawrence, Kansas.

Schram, F.R. 1986: *Crustacea.* Oxford University Press, New York, Oxford. 606 pp.

Selden, P.A. 1993: Arthropoda (Aglaspidida, Pycnogonida and Chelicerata). *In* Benton, M.J. (ed.): *The Fossil Record 2*, 297–320. London.

Shear, W.A. 1992: End of the 'Uniramia' taxon. *Nature 359*, 477–478.

Shu D.-g. 1990: *Cambrian and Lower Ordovician Bradoriida from Zhejiang, Hunan and Shaanxi Provinces.* 95 pp. Northwest University Press, Xian, Shaanxi. (In Chinese, with English summary.)

Shu D.-g., Geyer, G., Chen, L. & Zhang, X.-l. 1995: Redlichiacean trilobites with preserved soft-parts from the Lower Cambrian Chengjiang fauna (south China). *In* Geyer, G. & Landing, E. (eds.): Morocco '95 – the Lower–Middle Cambrian standard of western Gondwana, 203–241. *Beringeria Special Issue 2.*

Shu D.-g., Zhang X.-l. & Geyer, G. 1995: Anatomy and systematic affinities of the Lower Cambrian bivalved arthropod *Isoxys auritus*. *Alcheringa 19*, 333–342.

Simonetta, A. M. 1992: Wonderful life: the Burgess shale and the nature of history. [Book review.] *Ethology Ecology & Evolution 4*, 299–304.

Simonetta, A. M. & Delle Cave, L. 1975: The Cambrian non trilobite arthropods from the Burgess Shale of British Columbia. A study of their comparative morphology taxinomy and evolutionary significance. *Palaeontographica Italica 69*, 1–37.

Starobogatov, Ja.I. 1985: O sistema trilobitoobraznykh organizmov. *Byulleten' moskovskogo obshchestva ispytatelej prirody, Otdel geologicheskij 60*, 88–98.

Storch, V. & Welsch, U. 1991: *Systematische Zoologie.* 4th Ed. 731 pp. Fischer, Stuttgart.

Størmer, L. 1939: Studies on trilobite morphology. I. The thoracic appendages and their phylogenetic significance. *Norsk Geologisk Tidsskrift 19*, 143–273.

Størmer, L. 1944: On the relationships and phylogeny of fossil and recent Arachnomorpha. *Skrifter Utgitt av Det Norske Videnskaps-Akademi i Oslo. I. Matematisk-Naturvitenskapelig Klasse 1944 5.* 158 pp.

Størmer, L. 1955: Merostomata. *In* Moore, R.C. (ed.): *Treatise on Invertebrate Paleontology, P, Arthropoda 2*, P4–P41. Geological Society of America and University of Kansas Press, Lawrence, Kansas.

Størmer, L. 1959: Trilobitomorpha. Trilobitoidea. *In* Moore, R.C. (ed.): *Treatise on Invertebrate Paleontology, O, Arthropoda 1*, O22–O37. Geological Society of America and University of Kansas Press, Lawrence, Kansas.

Stürmer, W. & Bergström, J. 1973: New discoveries on trilobites by X-rays. *Paläontologische Zeitschrift 47*, 104–141.

Stürmer, W. & Bergström, J. 1976: The arthropods *Mimetaster* and *Vachonisia* from the Devonian Hunsrück Shale. *Paläontologische Zeitschrift 50*, 78–111.

Stürmer, W. & Bergström, J. 1978: The arthropod *Cheloniellon* from the Devonian Hunsrück Shale. *Paläontologische Zeitschrift 52*, 57–81.

Tasch, P. 1969: Branchiopoda. *In* Moore, R.C. (ed.): *Treatise on Invertebrate Paleontology, R, Arthropoda 4*, R128–R191. Geological Society of America and University of Kansas Press, Lawrence, Kansas.

Tiegs, O.W. & Manton, S.M. 1958: The evolution of the Arthropoda. *Biological Reviews 33*, 255–337.

Tollerton, V.P., Jr. 1989: Morphology, taxonomy, and classification of the order Eurypterida Burmeister, 1863. *Journal of Paleontology 63:5*, 642–657.

Walcott, C.D. 1905: Cambrian fauna of China. *Proceedings of the U.S. National Museum 29*, 1–106.

Walcott, C.D. 1911: Cambrian geology and paleontology. II. Middle Cambrian Merostomata. *Smithsonian Miscellaneous Collections 57:2*, 17–40.

Walcott, C.D. 1912: Middle Cambrian Branchiopoda, Malacostraca, Trilobita, and Merostomata. *Smithsonian Miscellaneous Collections 57:6*, 145–228.

Walossek, D. 1993: The Upper Cambrian *Rehbachiella* and the phylogeny of Branchiopoda and Crustacea. *Fossils and Strata 32*, 1–202.

Walossek, D. & Müller, K.J. 1990: Upper Cambrian stem-lineage crustaceans and their bearing upon the monophyletic origin of Crustacea and the position of *Agnostus*. *Lethaia 23*, 409–427.

Walossek, D. & Müller, K.J. 1992: The 'alum shale window' – contribution of 'orsten arthropods to the phylogeny of Crustacea. *Acta Zoologica 73*, 305–312.

Wang H.-z. (Wang H.C.) 1941: Note on the Chungitsun phosphate deposit, Kunyang, Yunnan. *Bulletin of the Geological Society of China 21:1*, 67–70.

Wang Y.-l. 1941: The age and cause of phosphorite formation in Yunnan. *Geological Review 6:1–2*, 73–94.

Whittington, H.B. 1971: Redescription of *Marrella splendens* (Trilobitoidea) from the Burgess Shale, Middle Cambrian, British Columbia. *Geological Survey of Canada, Bulletin 200*, 1–24.

Whittington, H.B. 1974: *Yohoia* Walcott and *Plenocaris* n.gen., arthropods from the Burgess shale, Middle Cambrian, British Columbia. *Geological Survey of Canada, Bulletin 231*, 1–27.

Whittington, H.B. 1975: Trilobites with appendages from the Middle Cambrian, Burgess Shale, British Columbia. *Fossils and Strata 4*, 97–136.

Whittington, H.B. 1977: The Middle Cambrian trilobite *Naraoia*, Burgess Shale, British Columbia. *Philosophical Transactions of the Royal Society, London, B 280*, 409–463.

Whittington, H.B. 1979: Early arthropods, their appendages and relationships. *In* M.R. House, M.R. (ed.): *The Origin of Major Invertebrate Groups*, 253–268. Academic Press, London.

Whittington, H.B. 1980: Exoskeleton, mould stage, appendage morphology and habits of the Middle Cambrian trilobite *Olenoides serratus*. *Palaeontology 23*, 171–204.

Whittington, H.B. 1985a: *Tegopelte gigas*, a second soft-bodied trilobite from the Burgess Shale, Middle Cambrian, British Columbia. *Journal of Palaeontology 59*, 1251–1274.

Whittington, H.B. 1985b: *The Burgess Shale*. 151 pp. Yale University Press, New Haven.

Whittington, H.B. & Briggs, D.E.G. 1985: The largest Cambrian animal, *Anomalocaris*, Burgess Shale, British Columbia. *Philosophical Transactions of the Royal Society, London, B309*, 569–609.

Wilson, G.D.F. 1992: Computerized analysis of crustacean relationships. *Acta Zoologica 73:5*, 383–389.

Zang W.-l. 1992: Sinian and Early Cambrian floras and biostratigraphy on the South China Platform. *Palaeontographica Abt. B, 224:4*, 75–119.

Zhang W.-t. 1987a: Earliest Cambrian trilobites from Yunnan. *Stratigraphy and Palaeontology of Systemic Boundaries in China, Precambrian–Cambrian Boundary 1*, 1–18. Nanjing University Publishing House, Nanjing.

Zhang W.-t. 1987b: Early Cambrian Chengjiang fauna and its trilobites. *Acta Palaeontologica Sinica 26:3*, 223–236. (In Chinese, with English summary.)

Zhang W.-t. & Hou X.-g. 1985: Preliminary notes on the occurrence of the unusual trilobite *Naraoia* in Asia. *Acta Palaeontologica Sinica 24:6*, 591–595. (In Chinese, with English summary.)

Zhou Z.-y. & Yuan J.-l. 1982: A tentative correlation of the Cambrian System in China with those in selected regions overseas. *Bulletin of the Nanjing Institute of Geology and Palaeontology, Academia Sinica, 1982:5*, 289–306. (In Chinese, with English abstract.)